Machinery

John Henry on Packaging, Machinery, Troubleshooting

A collection of writings from Food & Beverage Packaging Magazine

John R Henry CPP

www.changeover.com

www.foodandbeveragepackaging.com

On the cover
Oden PRO/FILL 3000
Automatic Liquid Filler
Used with permission
Oden Corporation
www.odencorp.com

Book Copyright © 2011 John R Henry

The articles contained in this book were originally published between 2001 and 2010 in Food & Beverage Packaging Magazine (Formerly Food and Drug Packaging) which holds the copyright.

Reprinted by permission. All rights reserved

ISBN-13: 978-1463579272
ISBN-10: 1463579276

DEDICATION

This book is dedicated to Alanys G & John M

They light my life.

ACKNOWLEDGMENTS

Jim Peters, Lisa Pierce and Pan Demetrekakes were my editors at Food & Beverage Packaging for these articles and columns. They taught me how to write and always made me look like a much better writer than I am. Thank you, I am forever in your debt.

Table of Contents

1 Six ways you can solve coding problems 1
2 Five common liquid filling problems—solved! 6
3 Solving five common labeling problems 11
4 Conveyor Basics: Keep the line running 16
5 Nine simple ways to cut changeover time 21
6 Five steps to better capping 26
7 How to solve the top 5 common cartoning problems 31
8 Making your Case with Corrugated 35
9 Focusing on 8 common vision system challenges 40
10 Solve common tamper-evident sealing problems 45
11 How to augment, troubleshoot your timing screws 51
12 Powder filling problems – Solved! 56
13 Putting the lid on 4 common cap sorting problems: 61
14 Seven ways to smoother bag/pouch packaging 66
15 Eight tips for more flexible packaging lines 71
16 In or Out? 77
17 Power at Your Demand 79
18 Machine vs. Materials: From Conflict to Cooperation 81
19 Changeparts versus machine adjustment 83
20 Lean changeover: It's as simple as 1-2-3 86
21 Minimize variation to maximize Quality 88
22 Speed! 90

23 RTM! (Read the Manual) 92

24 Make Machinery Manuals Easily Accessible................................95

25 New Beginnings..97

26 Sweat the Small Stuff 99

27 Evaluate the Cost/Benefit of Training 101

28 Metric versus Inch 103

29 Why and how to track OEE 105

30 OEE and changeover 107

31 Ten tips to simplify machine cleaning 109

32 Get Rid of Tools on the Line...111

33 New versus used 113

34 What you see is what they get 115

35 Know your costs 117

36 Don't cheap out 119

37 Get systems and people ready now 121

38 The Snake and the Seer 123

39 Lazy manufacturing or lean manufacturing?........................125

40 Rocket line starts aren't rocket science 127

41 Remember: Downtime means downturned pockets.........129

42 Fast forward: The packaging line of 2019 131

About the author 138

Introduction

When Jim Peters was editor of Food & Drug Packaging I submitted an article which was an overview of the various types of package coding systems. He liked it but suggested instead that I write an article on some of the common problems with coding systems as well as how to solve them. This became the first in a series of article on solving common problems with various classes of packaging machines. They were a lot of fun to write as they made me think in more depth about the machines I saw every day. They were particularly rewarding in the positive comments that they have generated over the years.

A few years later Lisa Pierce asked me to write a regular column on packaging, machinery and pretty much anything I thought would be of interest. Those were even more fun to write.

And perhaps the best of all, in 2009 for their special issue, Lisa asked me to write an article on what the packaging line of 2019 would look like. I like predictions about the future, mainly to laugh at how wrong they usually get it. Now the shoe is on my foot. Check back in a few years to see how close I was. No refunds if I wasn't.

Many of these articles and columns are available on the Web. Some in the magazine archives, others in other places. Unless one knows what to look for, they are not that easy to find. I hated the idea of them fading away, hence this book. I have corrected a few typos, punctuation and other formatting errors. Other than that, the words that follow are as they were originally published. I hope that you enjoy them. More importantly, I hope you find them useful.

It may appear that there were two different magazines. In December 2008, Food and Drug Packaging published its last issue. In January 2009, Food and Beverage Packaging published its first. They were the same magazine and same staff, just a different name and different editorial focus.

Feedback is always welcome. Let me hear from you

John R Henry CPP
johnhenry@changeover.com
www.changeover.com

John R Henry CPP

1 Six ways you can solve coding problems

Whatever coding technology you use, it has to produce a readable code with a high degree of reliability.

It must also be supportable by the people who are using it. In other words, it is not appropriate to use a complex technology that is beyond the capability of the plant personnel to operate and maintain.

But, no matter what technology you use, problems will arise.

Poor legibility is the most common problem. If the code isn't legible under its intended use, it's a problem. Legibility is subjective--it relates to size, contrast, character formation, fonts and other characteristics.

Case codes, for example, need to be very large so that they can be read from 6 to 10 feet away; they can be a dot-matrix format. The 40-micron high code on diamonds is visible only under a microscope, but the characters must be precisely formed.

Most applications will be somewhere in between those two extremes.

One trick to determine legibility is to look at 3s and 8s. And look at 5s and 6s. Is it easy to tell one from another? An incomplete 8 can look like a 3. A smeared 3 can look, like an 8. An improperly formed 5 may be indistinguishable from a 6.

A key question to ask: Will the code be read by people or by a vision system?

Machinery Matters

The human eye is very forgiving and often can read imperfect characters correctly. Vision systems, on the other hand, improve every year but are still somewhat finicky about what they will read. They will not tolerate much variation, and machine-readable codes must be more precise than codes to be read by humans.

Here are some of the most common coding problems and their solutions:

1. USE THE RIGHT APPLICATION PRESSURE

Improper application pressure results in deformed codes when you are using a contact coder such as a hot-stamp or ink coder.

If there is not enough pressure, the ink will not be transferred evenly, leaving gaps in the characters. This can easily make an 8 appear to be a 3. Too much pressure will deposit too much ink on the surface, causing, in the worst case, an unreadable blob of ink.

One variation of the problem occurs when the steel or rubber type is incorrectly mounted or worn, or when mismatched type is used.

In a contact coder, which uses either steel or rubber type, the surface height of all characters must be identical. If not, a high character may press too hard at the same time a low character is not pressing hard enough resulting in an uneven code.

Another contributor to improper application pressure is backing plate mounting. Backing plates, where used, must be square to the typeface. If not, one part of the code will be pressed against the material harder than another part. The plate must also be smooth and flat for the same reason.

2. USE THE RIGHT TEMPERATURE

Hot-stamp imprinters use a combination of heat and pressure to transfer ink from the ribbon to the product.

To some extent, heat and pressure offset each other. Increased heat may be used to compensate for insufficient pressure. This is an incorrect solution, though. There is an optimum setting and you need to make sure that the thermostat is properly adjusted and functioning. High temperatures will cause burning while low temperatures will not allow the ink to transfer properly.

3. CONTROL RIBBON USAGE

The ribbon used for hot stamping is relatively expensive. Care must be taken to assure that only the minimum amount of ribbon is used on each imprint. Most imprinters have an adjustment that controls the ribbon's advance on each cycle.

This should be set so that, looking at the used ribbon, one can see only a minimal amount of unused space between codes. Ribbon width should be specified to be only slightly wider than the code requirement.

4. USE THE RIGHT INK FOR THE RIGHT SUBSTRATE

There are hundreds of inks on the market. Coding ink must be carefully selected for compatibility with the substrate to be coded. The same rule applies for ribbons on hot-stamp or thermal transfer coders.

For example, ink that works very well on a paper label may not work at all on a plastic film label. It may be absorbed into the paper, but it just sits on the plastic's surface. Completely different types of ink are required for the two substrates.

Before making a final coding specification for a package, the best practice is to send samples of the product to be coded to the ink or ribbon manufacturer and ask for their recommendations.

Sometimes, the ink may be in contact with the product itself, and it must be compatible with the product.

For example, some inks can migrate through a plastic bottle wall. If the bottle contains a food or pharmaceutical, unacceptable contamination can result. This can be avoided by selecting ink approved for human consumption, by using a non-migrating ink or by selecting an entirely different coding method.

Here are two other substrate considerations to watch out for:

Label varnish. Many labels and cartons are coated with a varnish to give a glossy appearance for eye appeal. The varnish is non-absorbing and if ink is applied to it, the ink may often rub off or smear. If using a hot stamp, the ink may not even transfer from the ribbon.

One solution is to have the label printed with an unvarnished window where the code is applied. This window will not be noticeable but will allow proper ink application.

Mold release. Many plastic bottle manufacturers use a silicone-based mold release in the bottle manufacturing process. Residual release will remain on the bottle surface and must be removed via flaming or electrostatic treatment. If it is not, the mold release will prevent the ink from adhering properly.

5. MONITOR PRINTHEAD WEAR

Thermal transfer printers are similar to hot-stamp imprinters in that both use heat to transfer ink from a ribbon to a label.

Thermal transfer is an excellent technology but it has its own problems. As with hot stamp, these include temperature, pressure and ribbon/product compatibility. They also have problems with wear in the printhead.

This wear must be monitored and the head replaced as necessary. Routine cleaning of the head is a must as well. Care must be taken with threading the ribbon, as a wrinkled ribbon will result in unprinted areas on the label.

6. CONTROL HOW PACKAGES FEED TO THE CODER

The way packages are fed to the coder can affect code quality.

Non-contact coders and speed-controlled ink jets and dot matrix lasers work by scanning a series of vertical lines, which form the code's characters.

They do this at speeds ranging from very low (less than 5 feet per minute) to very high (more than 250 feet per minute). These printers work reliably at any speed in its range. However, the relative speed of the coder and product is critical.

If an ink jet is expecting the product to pass at a speed of 30 feet per minute, and it passes at 60 feet per minute, the spaces between the vertical lines of ink dots will be stretched horizontally. This will result in a series of dots, which look almost random. There will be no legible pattern.

If the product passes more slowly than the ink jet expects, the result will be the opposite; the space between the vertical lines of dots will shrink until the code becomes an unsightly blob.

One way to counteract this problem is to use an encoder to monitor the product speed. This is commonly mounted on the conveyor's drive shaft. The encoder is integrated with the printer and provides feedback on the actual speed, resetting the printer as necessary.

Backlog control is another aspect of package handling at the coder. Products must not be allowed to back up in front of the coder. If the downstream machine is running slowly or if there is a jam, there needs to be a sensor which will stop feeding product to the coder. If products back up in front of the coder, the code will be just a blob.

ANSWERS TO THE INK JET MESS

A common complaint against ink jets is their mess. They will frequently drip, or overspray Occasionally they will be triggered inadvertently with no product in front of them. Some plants place black floor tiles under and around the ink jet printer. While this does not eliminate the mess, at least it hides it to some extent.

One way to reduce the messiness is to place a shroud with extractor around the point at coding. This will remove the ink droplets tram the air before they get on the machinery or the floor.

Another solution is one implemented by Luciano Packaging Technologies on a cartoner. Their solution was to move the carton so that it overhangs the conveyor. The ink jet nozzle is mounted above the carton and sprays down. A stainless steel tray under the carton then collects over-spray and drips. The tray is removable for ease at clean-up.

A primer on coding options is on our Internet site. It covers contact coding such as debossing, roller coders and hot stamping. It also details non-contact options such as ink jet and laser coding.

This article originally appeared in the September 2001 issue of Food & Drug Packaging magazine

2 Five common liquid filling problems—solved!

The magic isn't pulling a rabbit out of a hat. The real trick is getting it there in the first place. Packagers face a similar problem getting their product into the container.

Here are some common liquid filling problems and tips on solving them:

1. FOAMING

Foaming can cause the product to overflow resulting in both mess and an underfilled container. Foaming is invariably caused by entrainment of air in the product, generally from one of three sources.

Air can be introduced into the product prior to filling by overly vigorous or improper agitation during the manufacturing process.

It can also be introduced when the product is pumped too vigorously into the filler reservoir. At this stage, if air entrainment is the problem, a review of the product's mixing and handling is in order.

A leak in the filler's inlet side and a negative suction head can combine introducing air into the product. Best practice dictates maintaining the infeed plumbing under a positive head at all times. Gravity is generally preferred as it is constant and requires no controls. Elevating the product reservoir above the filler, pressurizing it with an air or nitrogen blanket or both can do this.

If positive pressure is used, more complexity and pressure is introduced into the system, which will cause problems somewhere down the road. Note that I said will. There is no escape from Murphy's law.

Finally, air can be entrained if the product is blasted into the container at high velocities. The turbulence and roiling in the container will mix air into the product.

There are several cures for this. The most obvious one is to slow the filler. But this is usually not feasible due to production requirements. An alternative is using a larger filling nozzle diameter to decrease fluid velocity while maintaining the same flow rate.

We've all seen how pouring beer down the side of a glass reduces foam. This same idea holds just as true in a bottle.

A side-shooting nozzle is available which dispenses the product along the side rather than straight down. Dispensing the product down the sidewall gives a gentler flow and reduces foaming.

Subsurface filling may be the answer in some cases. In subsurface filling the nozzle is lowered to the bottom of the container at the beginning of the fill cycle. As the product is dispensed, it covers the tip of the nozzle reducing turbulence. One drawback to subsurface filling is that the outside of the nozzle is now wetted and will drip when withdrawn.

The key to combating foam is keeping air out of the product and the key to keeping air out is gentle handling. Gentle handling can also reduce splashing.

2. SPLASHING

Splashing caused by product velocity is often a problem in open mouth containers such as jars. The solutions for splashing are similar to those mentioned for foaming. By using larger diameter nozzles, side-shooting nozzles or subsurface filling, product velocity can be reduced.

3. VOLUME CONTROL

Volume is the most critical parameter of filling. If the actual volume is less than label claim, the customer is cheated. If volume is more, the manufacturer is giving away product.

Machinery Matters

Even seemingly insignificant amounts of giveaway add up. A 12-ounce bottle running at 250 bottles per minute with an average overfill of 0.1 percent will, in the course of a shift, result in the loss of almost 1,000 bottles of product.

There are two concerns with volume. One is the actual, measured volume and the other is the perceived volume. Although related, they are not always the same.

Some containers, mainly glass, will vary in their internal volume. When this occurs, two bottles with identical volumes of product will have different liquid levels. If consumers see bottles on the shelf with apparently varying volume, they will be turned off.

An alternative is to use a neckband to hide the actual product level. Or a "level" or "cosmetic" filler can be used. These fill to a level rather than a specific volume.

Level fillers are generally pretty straightforward and simple but can still have problems.

One common problem with level fillers relying on overflow occurs when the filling cycle is too short. In this case, the product never reaches the overflow or shutoff level resulting in short and uneven fills.

Many level fillers rely on a seal between the filling nozzle and the neck of the bottle. If this seal leaks, product will overfill and may overflow.

Other styles of level fillers use electronic or mechanical sensors to determine liquid level and control the flow. If the sensor is damaged or improperly adjusted, variation will occur.

Volumetric fillers measure fill volume externally rather than via the bottle. They come in a variety of different architectures.

With piston fillers, the volume dispensed is a function of the piston diameter (bore) and the stroke length. A common problem occurring with piston fillers is play in the drive train. This play causes the stroke of the piston to fluctuate, varying volume. When this occurs the solution is to replace all drive components, shafts, linkages and bearings to bring the system back to its original specifications.

Piston fillers require valves to control the product flow on intake and discharge stroke. If these valves leak or are improperly timed, air can be sucked into the nozzle or product can be pumped back up the infeed. In small amounts, this may not be directly noticeable but it will still cause variations in fill volume. The solution is to adjust, repair or replace the leaking valves.

Filling pistons have both an intake and discharge stroke. The infeed stroke draws a measured amount of product into the piston. If the stroke is too fast for the product flow characteristics, it may not fill completely.

When the piston discharges, a short fill will occur. This problem can be solved by increasing product head, using larger diameter tubing, slowing the pump speed or a combination of all of the above.

Some volumetric fillers are sensitive to infeed conditions. Two important conditions are infeed head and product viscosity.

If the filler is connected directly to a compounding tank the head will decrease as the tank is drawn down. This decreasing head may result in a reduced fill volume. The way to correct this is to add a small reservoir between the main tank and the filler. The reservoir can be maintained at a consistent level throughout the filling process.

Temperature fluctuations will alter the viscosity of products like shampoos or syrups. Performance of the filler will vary accordingly. The solution for this problem is to use a temperature-controlled tank or to cool or warm the product to room temperature prior to setting up the filler.

4. DRIPPING NOZZLES

Some people think fillers are supposed to drip. They are not. In volumetric fillers, leaky valves usually cause dripping. If the product reservoir is elevated product will naturally try to flow through the filler and out the nozzle. It is stopped from flowing by a shutoff valve.

Different manufacturers use different types of valves but most can leak under the right circumstances. Common problems include worn seals and seating surfaces, improper adjustment and product buildup.

The cure is simple. Make sure the correct valve has been selected for the job and then make sure it is maintained and set properly.

Machinery Matters

One way NOT to solve the problem is by lowering the product reservoir. This may stop dripping but will cause product to backflow, bringing air into the filler and causing underfills.

Dripping also occurs when the nozzle is in contact with the product. Product contact may be the result of subsurface filling or foaming. In these cases the dripping comes from product remaining on the outside of the nozzle.

There are a couple of solutions to dripping. The simplest is to have a small drip pan that goes between the container neck and nozzle to catch the drip. Another is to use a vacuum aspirated nozzle. This nozzle contains a secondary path under vacuum. As the product tries to drip, it is caught by the vacuum line and carried to a catch tank.

5. STRINGING

When pouring honey, a thread or "string" of honey remains after pouring has stopped which takes a moment to break. In filling, moments are precious and there is no time to wait for the string to break naturally. On the other hand, if the container is moved before the string breaks, product will wind up on the outside of the bottle.

To control stringing, nozzle design is critical. Different designs are available. Some use a mechanical device like a knife blade to physically cut the string. Others have an air channel in the nozzle, which breaks the string with a puff of compressed air.

USING THE RIGHT FILLER

Probably the best way to avoid filler problems is to select the most appropriate machine for the job. There are at least 15 distinct filling architectures commercially available--such as piston, gear pump, vacuum, net-mass and time-gravity. Within each broad architecture, equipment manufacturers will put different spins on execution. It's safe to say that almost all of these can work properly in certain applications. It's also safe to say that each will work poorly in applications.

The proper marriage of filler, product and container, along with the up and downstream equipment will prevent a lot of gray hair and sleepless nights.

This article originally appeared in the December 2001 issue of Food & Drug Packaging magazine

3 Solving five common labeling problems

Regardless of a label's material and adhesive, it has to be both aesthetic and functional. The label must be undamaged, smooth and positioned correctly during application.

A number of problems can occur in labeling. Here are five of the more common ones and some ideas on how to solve them.

1. CONTAINER CONTROL

One common reason for misapplied labels is a lack of container control.

Most dispensing heads will dispense labels with a high degree of repeatability. If the product is not under control, label placement will vary.

For example, placing a front and back label on an oval bottle can be a challenge. The long axis of the container must be perfectly parallel with flow.

When the labeler is set up, the bottle is straight. But, during production, it may turn slightly. This may not be noticeable without careful observation and will result in one label closer to the leading edge of the container and the other closer to the trailing edge.

Several factors can cause this including:

> A timing screw with incorrectly cut pockets. Or, if the screw is damaged, the product will not be aligned.

Machinery Matters

A loose product in the pocket can cause twisting. You must ensure that the guide rail and screw(s) are properly adjusted, holding the product tightly.

When the product enters the labeling section, make sure it stays aligned. Often the product is held in orientation by a top hold-down belt. If this belt is loose or worn, the product will not be firmly held, and movement can occur.

In rotary labelers, the product is placed on a pad and held down by a top piece. A mechanical cam or a servo-motor then rotates it to the proper orientation. If these are not adjusted, the container won't be positioned correctly.

2. SKEW

When the axes of the label and product are not parallel to each other, it is called "skew." It is especially noticeable on round bottles with full wrap labels where the label ends will not match up.

Skew can occur when the product or the labeling head are out of plumb. When relating to machine set-up, it can be a permanent problem. Or, it may be an intermittent problem related to product movement or machine looseness.

You should verify that the labeling machine is sufficiently rigid. Some low-cost labelers are made of lighter gauge materials, which are prone to vibration and misalignment. Bracing the machine may steady the head and resolve the problem.

Also, imperfect molding of plastic and glass bottles can result in them not being perpendicular. When this occurs, skew can vary all over the place, depending on which way the bottle is leaning when the label is applied.

Placing the bottle on a flat surface and applying a square at four points around the circumference can identify imperfections in bottles.

A bottle with a bulged or non-flat bottom can also be subject to skewing. If this is the case, you may have to go back to the supplier and insist they get their process under better control.

3. WRINKLES

Wrinkled labels can result from several causes. Perhaps the most common occurs when the label is wiped onto the product when the label dispensing and product speeds are not exactly synchronized. If the product is moving more slowly, the label will wrinkle as it tries to push the product.

In some cases, you may have to consider a different kind of a label applicator. For example, a vacuum grid, roll-on applicator will pre-dispense the label onto a plate where it is held in place by vacuum.

When the product passes, it makes contact with the end of the label and pulls it from the grid. A roller or brush wipes the label down as it is pulled.

Air-blow or tamp-on applicators can also delink label dispense and product speeds. Both techniques dispense the label onto the vacuum grid or pad to await the product.

When in position, the label is either blown or pressed onto the container. Air blow is especially useful for smaller labels at high speeds since it is noncontact. I once used a low cost air-blow system to apply a 3/4-inch round label to can lids at over 500 cans per minute (cpm).

Container design can also cause wrinkling. For the most part, if the label is flat it can be wrapped around a simple curve such as an oval bottle. But, it cannot conform to a complex curve such as a keg shaped bottle.

When the label tries to conform to the container's unusual curve, it will wrinkle as it tries to lie flat. This is usually a problem of product design, but can also be caused by other factors.

A container that is designed to be straight-sided may deform due to improper storage. A small amount of deformation, not even noticeable to the eye, will cause wrinkling when the label is applied.

Some containers, such as plastic, half-gallon milk jugs, will deform in production. Even though they are straight sided when empty, containers with flimsy sidewalls bulge when filled. The solution here is to label the bottle before filling.

Wrinkling can also be caused by poor adhesion of the label to the product. There are more than 300 types of adhesives used for labeling, each specifically formulated for certain conditions. A label designed for a plastic bottle may fall off of a carton.

Adhesives for labeling cold products may not work at all for hot products. Improper selection of the adhesive for the product and conditions is the most common reason for poor adhesion.

When labeling a cold product, condensation may be present. If so, the moisture will prevent label adherence. To remove moisture, blow warm, dry air on the product immediately prior to labeling.

If the compressed air used to assist label dispensing contains oil or moisture, the adhesive can be compromised. A good moisture/oil removal system can help solve this problem.

If mold release is used in bottle manufacturing, it will leave a thin film on the product, which interferes with adhesion. In these cases, the product surface needs to be treated prior to labeling. Flame and electrostatic discharge are two common treatments and are usually done by the supplier.

4. TAMP PAD PROBLEMS

Tamp applicators consisting of a pad (usually rubber) mounted on an air cylinder are popular because they work well in many applications. But when the vacuum holding the label on the pad is insufficient, the label will slip or even fall off. Venturi vacuum generators will often accumulate dirt inside. Opening it and cleaning may resolve the problem.

Leaky or pinched vacuum hoses can cause similar problems.

When the tamp-pad wears it may do so unevenly. If the pad is not absolutely flat, air will leak between the label and the vacuum holes. When this happens, the vacuum may not hold the label in position.

Set up of tamp applicators seems counter-intuitive. Logic dictates that the pad should go slightly above the labeler peeler plate. Logical, perhaps, but incorrect.

As the pad comes down--if the pad is above the peeler plate--the edge of the label will strike the peeler plate and, especially with smaller labels, get pulled out of position or pulled off the pad completely. Be sure to set the pad to the machine manufacturer's recommendations, which usually dictate settings slightly below the peeler plate.

5. SENSOR PROBLEMS

Roll-fed labelers generally use a photoelectric sensor to control where the label stops for dispensing. This sensor must be properly set to stop the label in the correct position.

Be sure to set the sensor to the machine manufacturer's specifications. And, be sure the label and the sensor are compatible.

In most cases, the sensor will sense either the gap between the labels or registration marks. If the sensor cannot detect this, it will not work properly.

The linear position of the label on the product is usually determined by the position of the trigger sensor relative to the labeling head.

Setting this position correctly can be tricky and time consuming. To reduce the set-up, consider mounting multiple trigger sensors.

A separate one is mounted in the correct position for each product. A selector switch is used to select the proper one. This not only saves time and ensures a perfect set-up, it may also allow the set-up to be performed by an operator rather than a mechanic.

The label is usually the primary means by which a manufacturer communicates information to the customer. Obviously, a missing or illegible label conveys no information.

Getting the label right is of primary importance. A properly placed, smooth label conveys the message that the product is quality.

Many of the problems, as mentioned above, are relatively simple to solve. Proper labeling is not difficult if all the components are correct.

Proper labeling requires meticulous attention to detail during set-up and operation. Small errors or incorrect adjustments may seem trivial but will be magnified during operation, resulting in poorly applied labels.

This article originally appeared in the January 2002 issue of Food & Drug Packaging magazine

4 Conveyor Basics: Keep the line running

We usually pay nowhere near enough attention to conveyors yet they have been called "Intelligent bridges between islands of automation" and they are a key component of any packaging line. If the bottle cannot get from the filler to the capper, it doesn't matter at all how wonderful and efficient either of them are.

Some things to consider when choosing a conveyor system for your packaging line are its construction, drives, transfers and backpressure control.

Construction

Flat top chain is the most commonly used style of conveyor in packaging lines and will be the focus of this article. Belt, roller and other styles are used but more frequently in end of line applications.

Flat top chain consists of a series of links, each an inch or so long, pinned together. The top of the chain is a plastic or metal "flight" which can be from 1" to 12" wide. Mat-style chain is a variation that allows almost infinite width.

Flat top chain has the advantage of readily permitting slip of the product when it is queuing for a machine infeed. This allows the conveyor to be run slightly faster than necessary to keep the products bunched up and under a slight backpressure which helps assure smooth feeding.

Flat top chains are available in side flexing formats which easily allow them to go around curves. While roller and belt conveyors can have curves, they are more difficult.

There are several generic styles of for flat top chain conveyors. The most common is generally referred to as "channel frame" construction. It consists of two vertical "C" shaped sections, held apart by spacers with a sprocket at each end and. The chain rides on top of the frame. The return chain can be either suspended from the lower "C" section or may ride on rollers or guides.

Another style is "sanitary" construction. A sanitary style conveyor consists of an inverted "U" shaped box section, typically about 6" deep and slightly wider than the chain. A pair of rails mounted an inch or so above the surface of the frame provides a running surface for the chain. The return chain is normally enclosed within the lower portion of the frame. The great advantage of this construction comes when packaging liquids, or powders. When a spill occurs, it will run through the chain but will not get inside the conveyor frame itself. This keeps the conveyor clean and avoids excessive cleanup times.

A third style is "open frame" construction and its use seems confined mainly to the beverage industry. The frame is constructed of angles and the frame sides are left open. This allows complete hose down without the need to remove the chain.

Conveyors are available in stainless steel, painted steel, plastic, aluminum and other materials. Although stainless steel will add 5-10% to the initial purchase price when compared to painted steel, this is generally a good investment. Regardless of the quality of the initial painted finish, in several years it will be chipped, scratched or dirty and will require repainting. The first repainting will generally obviate any initial cost savings. Additionally, a stainless steel conveyor just looks better and is easier to clean and maintain.

When designing conveyors, keep the sections as long as possible. A 10' long conveyor of a single piece will be more rigid and will give fewer problems than the same conveyor made up of 2-5' sections.

Drives

A critical conveyor design element is the drive system. Conveyors are commonly driven via an electric motor, through a gearbox and then to the drive sprocket via roller chain. V-belts can be used but may have problems of particulate generation and slippage. Conveyors can also be driven via a combination motor/right angle speed reducer mounted directly to the sprocket drive shaft. Drives can be either fixed or variable speed. Fixed speed drives will save a few dollars on initial costs but do not allow flexibility to fine-tune the speed for optimal operation.

Machinery Matters

Several companies manufacture drives which use internal devices to mechanically vary the output speed while the motor speed remains constant. If V-belts are used, spring loaded pulleys and a sliding motor base can be used to adjust speed. While mechanical speed adjusters can be reliable, they are mechanical devices and add complexity.

Conveyor speeds can also be controlled electronically. AC motors can be controlled via variable frequency controller. On smaller drives, under about 1HP, DC motors with SCR controllers are common. The advantage to electronic control is that remote or automatic speed control is possible.

Dual speed drives can prove useful in some cases. The conveyor will run at normal speed unless an excessive backlog occurs at which point it will automatically slow down to reduce backpressure until the backlog occurs.

Conveyors need to run at the right speed yet many, perhaps even most, conveyors have no way to determine what the speed is. A good practice is to have a speed indicator or tachometer on each conveyor to assure that it is properly set. Several manufacturers offer SCR speed controllers which include a built in tachometer and feedback loop. The conveyor is adjusted to the proper speed via the tachometer and the controller than varies output to compensate for any speed variations that might occur due to changing loads.

Transfers

Most lines will have more than one conveyor which makes transfers from one conveyor to another a fact of life. Transfers probably cause more problems than any other single facet of conveyor operation. The two most common transfers are end-to-end and side-to-side, also called end or side transfers.

In an end transfer, the product flows in a straight line from one conveyor to the next. Conveyors are butted end to end but, due to the radius of the sprockets, there will be a gap. This gap will be about 6 inches on flat top chain conveyors and must be bridged by a deadplate.

Deadplates cause a couple of problems. As the product moves over the lip of the deadplate, it can catch and tip, especially if the product is unstable to begin with. The second problem is that once the product is on the deadplate, it will stay there until the next several products come along and push it off. In the case of small or lightweight products, a considerable backlog may be required. Using a deadplate with rollers on it can alleviate this but these may not work with smaller products due to the pitch between rollers.

Side transfers avoid these problems by overlapping conveyors and using the guide rails to gently nudge the product from one conveyor to the next. Where side transfers are to be used, it is important to overlap enough to allow for gentle side motion. It is also important to build the conveyors without protruding hardware to allow them to nestle closely together. Ideally, they should be built so that the two belts touch with no gap.

A drawback to side transfers is that as product width changes, the radii of the inner and outer will not be parallel. An elegant way around this problem is the use of an "S" transfer which combines the best of both end and side transfer. In an "S" transfer, the conveyor ends have offsets so that they can overlap while leaving the product to flow in a straight line.

Backpressure Control

No discussion of conveyors would be complete without addressing backpressure. Generally, some backpressure is a good thing as it assures that the product is moved positively into each succeeding machine. Too much backpressure will be as bad or worse than too little as it can cause damaged products, jams, especially at curves and machine infeeds, and overloading of the conveyor system. There are a number of ways to control backpressure. Proper selection of conveyor chain, in most cases, will be sufficient. Chains are available in a variety of materials with a range of friction coefficients. Choose one that is "sticky" enough for the purpose while still allowing the chain to slide under the product as necessary. In extreme cases, zero backpressure chain can be used. The flights of these chains have small rollers on which the product rides, resulting in almost zero friction.

Many breweries and beverage plants will continuously spray water containing a lubricant on their conveyors. This not only keeps the chain clean, eliminating sticky deposits, it also reduces the friction between chain and bottle.

Conveyor zoning and backlog controls are another means of controlling backpressure. Zones are established throughout the conveyor system and a system of sensors, controls and stops will maintain the proper number of products in a zone at any given time.

Conclusion

Machinery Matters

The above has merely skimmed the surface of some of the issues involved in conveyor design. It is important that in designing any packaging line, as much thought must go into the conveyor design as any other aspect.

This article originally appeared in the March 2002 issue of Food & Drug Packaging magazine

John R Henry CPP

5 Nine simple ways to cut changeover time

Every company will have its own cost calculations for downtime due to changeover. These costs will range from hundreds of dollars to thousands of dollars per hour. One thing is universal though, changeover is always expensive and a program to reduce it is easily justified. Significant reductions can usually be achieved in a short period of time for a relatively small expense.

Changeover is the total process of changing a machine or line from running one product to another. The changeover clock begins when a line slows down at the end of one product run and doesn't stop until the next product is up and running at normal speed and efficiency. It may help to think of the clock in terms of dollars instead of seconds. One of my clients has a display on their lines that does just that. When the line shuts down, a display starts clicking off 73 cents every second ($2,736 per hour). It adds up quickly.

One major component of changeover time is start-up (sometimes called "ramp-up" or "run-up") time. It is the time spent tweaking the line after it has been restarted but before it has settled down. It is characterized by frequent starts and stops and high levels of rejected product. Start up time occurs because the line was not set up correctly or because of variation in the product or the components from lot to lot. Improving changeover will not directly affect product/component variation. However, if variation in the set-up is eliminated, it will focus a spotlight on it and make it's elimination a priority.

Here are some practical ideas that will reduce changeover times:

Machinery Matters

1. Standardize

A look at a road map of the U.S. will quickly show that there are a seemingly infinite combination of roads that can be taken to drive from Seattle to Miami. Only one combination is the best route.

If there are three mechanics performing changeover, they are probably doing it at least four different ways. Perhaps more. Obviously, they can't all be the best. The changeover must be carefully analyzed to determine the optimum process. This optimum process must then be documented in an SOP (Standard Operating Procedure) and the set-up people thoroughly trained in its practice. This is not the end. Changeovers must be monitored to assure that the SOP is being followed. As improvements are discovered, they are included and put into practice by all.

2. Measurement, repeatability

A common problem with changeover is that settings are done by eye and by feel. When this is the case, the adjustments will vary from mechanic to mechanic and even from day to day by the same mechanic. All adjustments must be quantifiable. That is, there must be a positive indication of set point such as a digital indicator or scale. Digital indicators are generally preferred as they are less subject to interpretation. Where scales are used, they should be marked in one tenth of an inch or in metric rather than in one-sixteenths. Most people have an easier time understanding decimal measurements than fractions.

Gauges are another possibility but are a tool and should be avoided whenever there is a better alternative available. Where gauges are used, they should be permanently attached. This can be with either a keeper wire or by permanently mounting them on a hinge or pin so that they can swing out of the way when not in use, but be readily available when needed.

3. Wide cams

One of my clients fills a powder on a vertical pouching machine. A knife is adjusted horizontally to cut the pouch to the proper length. When the knife is adjusted, it's actuating cam must be moved to match. They eliminated this adjustment by fabricating a wider cam. Now the knife may be moved from the shortest to the longest package with no need to move the cam.

This not only eliminated the step of adjusting the cam, but eliminated the need to remove and replace a side panel in the process.

4. Switched sensors

In many machines it's necessary to adjust photoelectric sensors or other sensors for varying product sizes. This may involve sensitivity settings as well as physical movement. An easy way to improve this is to use permanently installed, multiple sensors and a selector switch. Instead of adjusting the sensor, the appropriate, previously set, sensor is selected. This not only avoids the time and problems caused by sensor adjustment, it may also eliminate the need for a mechanic as an operator may be permitted to move the selector switch, This same technique will also work where there is an air cylinder that needs to be repositioned. Multiple cylinders with a multi-position valve can he used to eliminate adjustment.

5. Multi-tube connectors

Some machines look like they are covered with spaghetti. Don't eat it, though, they are the pneumatic control tubing. When it needs to be disconnected for changeover, the possibility of error during reconnection exists. Color coding the tubing is a must but is still not foolproof. Multiple connectors are commercially available which allow up to 10 1/4-inch tubes to be simultaneously connected via a single fitting.

6. Changepart storage

Are you spending valuable changeover time looking for changeparts? This is a procedural step that's easy to correct. A properly organized cart will improve changeovers in three ways:

> Good storage will prevent damage to fragile changeparts.

> A well designed cart will provide a quick visual confirmation that all parts are available prior to beginning the changeover.

> The cart can be brought to or near the machine prior to shutting down from the previous run. This prevents lost time chasing parts.

7. Tool Elimination

Machinery Matters

Eliminate the use of tools as much as you can. Time is often lost looking for the proper tool. Even worse, time is not lost and the improper tool is used. We've all seen nuts rounded off by the use of pliers or screwdrivers ground to a chisel point which strips the screw slot.

Another reason to eliminate tools is that plant policy may permit only mechanics to make adjustments requiring tools while allowing operators to make tool-less adjustments.

There are a number of techniques to eliminate tools:

Hand knobs and hand levers--Hand knobs should be used whenever the fastener needs to be removed and handle-less used when they only need to be loosened. This is not only good ergonomics. It also provides a memory tool to prevent levers from being removed unnecessarily.

Toggle clamps are available in a variety of sizes and styles.

Eliminate the use of tools as much as you can.

They're handy for holding things in place. One caveat is that, if used, they must be used so that they're fail-safe. That is, if they open inadvertently, catastrophic failure should not be the result.

Pins are available in a variety of sizes and styles and can often be used to lock components in place. I'm particularly partial to the positive locking pins with a pushbutton in the end.

8. Power tools

While tools should be avoided, there are instances when a power tool can not only help reduce changeover time, it can also make a better quality changeover and provide health benefits. For example, one manufacturer's blister packaging machines have about 30 points of adjustment. Normally, these are adjusted by the mechanic turning a small crank on each one to reach the proper set point. The machine maker now provides a power screwdriver with the charger permanently mounted on the control panel. A bit mates with the adjusting points.

One of my recent projects was reducing the changeover time on a tablet press. Fifty-five die bolts had to be removed and exactly re-torqued. A power torque wrench sped the process considerably while giving superior control of torque accuracy. A benefit even more important than the reduced changeover is that it reduces the ergonomic risk from repetitive wrist motions.

9. Unitized spacer blocks

Laners, such as in a case packer infeed, often have a number of spacer blocks that separate the lanes. These spacers can get lost or damaged and there might be three to four sets with as many as six blocks spacers to a set. Reduce the number of parts to be handled by machining a spacer set out of a block of plastic. Even better, machine two or more sets of spacers in each block. Thus, when changing from one size to another, it may only be necessary to flip the block over for the new pattern.

Conclusion

Albert Einstein reportedly once said "Imagination is more important than knowledge". As the above examples show, there are a number of simple and low-cost ways that you can reduce changeover times. The key is to look at the process and then think about what can be done better or faster.

One way to do this is to break the changeover down into the smallest possible elements and then brainstorm each one to find ways to eliminate, externalize or improve them. The mind is an amazing thing and you may be surprised at the ideas it will come up with when allowed. Never be scared of wild ideas. The wild idea of today is the conventional wisdom of tomorrow.

This article originally appeared in the June 2002 issue of Food & Drug Packaging magazine

6 Five steps to better capping

Consumers don't notice caps most of the time and that's the way it should be. When they do notice them, it's usually because they have a complaint. Criticisms generally relate to the screw cap being too tight (preventing easy removal) or too loose (resulting in leakage). Let's look at causes and solutions to both problems.

1. Torque

Torque is the rotational force required to either seat a cap (on-torque) or to remove it (off-torque). Generally, there is little or no torque involved in cap application or removal except at the last (or first, depending on which way you are going) eighth of a turn or less.

In automated capping, we are controlling on-torque hoping to control off-torque. Generally, for any given off-torque, a greater amount on-torque will be required. Torque is a critical packaging parameter. If too much torque is applied the cap will not only be difficult for the consumer to remove, it will also cause excess stress in the cap, which can result in cracking or thread stripping. Insufficient torque can result in leakage. This leakage may not always exist when the container leaves the plant but may develop as transit vibration causes the cap to loosen.

On-torque measurement:

The good news is that for a given container/cap combination, the relationship between on- and off-torque will be fairly constant. Several capper manufacturers supply torque monitors for their machines, which allow the on-torque applied to each cap to be measured. These systems will collect the data and provide for rejection of individual containers that are out of spec as well as collect statistical data to allow trend analysis.

The bad news is that even if the on-torque is controlled, there can still be variations in the off-torque and this is the parameter that really matters. And it is still necessary to monitor the off-torque.

Off-torque measurement:

Off-torque must be measured by use of a torque meter. Usually this is a small mechanical or electronic device consisting of a bottle clamp and a readout. The operator places the bottle in the clamp, grips the cap and twists. The torque meter will indicate the highest torque before the cap breaks loose. This sounds pretty straightforward but in reality is more complex. The off-torque and the indicated reading can vary considerably depending on operator technique. An operator applying a slow and steady twisting motion will generally indicate a higher reading than one when the cap is given a quick, sharp, twist. For reliable readings, operators must be carefully trained on how to use the meter. The way in which the cap is gripped can also have an effect on the reading. One way to control that variable is to use a chuck with a "T" handle to grip the cap.

The best solution is to use an automated torque meter. These include the typical torque meter but also include a motorized chuck. This removes all operator technique as a variable in off-torque measurement.

Torque control:

Most screw cappers use either a chuck or friction wheels to apply torque. Chucks come in several styles but all come down over the cap and rotate on the centerline of the cap axis. In a friction wheel capper, a pair of wheels rotate on either side of the cap. As the cap passes between them, they contact the side of the cap and spin it down.

Both chuck and wheel cappers usually use clutches to control the amount of torque applied. Clutches can be magnetic, pneumatic or mechanical. Whichever it is, they must be correctly set.

It would seem that if the clutch is set correctly, the correct on-torque should result but this is not always the case. If the chuck or the wheels should slip against the cap, the correct torque will never be achieved. In some instances, operators will actually try to control torque via slippage rather than the clutch but this will almost always result in poor control. When a clutch is working correctly, it should be easy to tell by visual inspection.

When capping, the chuck or wheels will noticeably stop for a moment on each cycle. At higher speeds, it may be necessary to use a strobe to see this. If they do not stop, they are either slipping or the cap is not being run down tight. Additional signs of slippage are chafing or wear marks on the cap, excessive wear of the chuck or wheels and particulate matter in the capping area.

The speed of the chuck or wheels can also have an impact on on-torque. A chuck is a mass of spinning metal, which has momentum. This momentum will continue to apply force to the cap even after the chuck releases. While it cannot be eliminated, it can be controlled. Minimizing the mass of the chuck helps but the real key is to find an optimum speed and maintain it.

Friction wheels have a minimal amount of time in contact with the cap. It is critical that they are spinning fast enough to apply the maximum desired torque. The comment about momentum also applies to the wheels. Again, the solution is to find the optimal speed and be sure that the capper is always set to it.

2. Cocking

If the cap is not correctly applied to the container, it can go on crooked, resulting in cross threading and improper sealing. The most positive way to apply a cap is to place it straight down over the container neck. This assures that it is square. This is the only way to apply some caps, especially those that have a dip tube or other feature extending below the cap skirt. Some cappers will have the chuck pick up the cap from a fixture and then place and rotate it with the same motion. This is the most positive technique.

While the above provides positive control, it also makes for additional mechanical complexity. Many cappers will use the container to place the cap. The cap is fed down a chute and held in an escapement at an angle over the centerline of the container path. The leading edge of the container neck then makes contact with the inside leading edge of the cap, pulling it out of the escapement. As it does so, the cap is guided to fall over the neck. A chuck or wheels then spin it down. One area of concern is to be sure that the cap guides are appropriately adjusted to control the cap and assure that it falls square over the bottle neck.

3. Container control

Containers must always be under control on any packaging line. The moment when the container and cap come together is especially critical. Containers must be stabilized and the neck centered under the cap. Note that it is the neck that is most critical.

Containers will sometimes wobble or be slightly off plumb. If necessary, add guides to assure that the neck is where it needs to be, especially at cap application.

Containers must not be allowed to rotate while the cap is being torqued. There are a number of techniques used to prevent rotation such as clamps, fixtures, starwheels or side belts. Where adjustable clamps or belts are used, it is important not to apply so much pressure as to deform the container. Any container deformation that occurs when the cap is tightened will remain and will cause problems with other downstream operations such as labeling.

4. Neck supports

Capping will apply downward force on the container. If the container is soft (a half-gallon milk jug is a good example) it may not be possible to apply enough force to get a good seal. Neck supports can help resolve this problem. In a rotary capper, they would be a starwheel that captures the container under a reinforcing ring under the neck. Inline cappers do the same thing by providing a pair of fixed guide rails to support the neck.

5. Cap-container tolerance

Initial capper settings are usually established for a particular cap and container combination. As long as both remain constant, all should be well. But they may not remain constant. Purchasing may get a good deal from another cap supplier without telling anyone in manufacturing. The caps may seem identical but may have a slight difference that will drive the line mechanics crazy trying to get it to run correctly.

One example might be the balance point, which could cause the cap feeder to jam or just not run as fast as normal. Even color can have an effect. One milk bottler changed from a blue to a red cap and found that the new cap was just a bit slipperier causing problems with achieving the correct torque.

Machinery Matters

Even if the same type of caps are used, they can change dimensionally as they age. "Fresh" caps may run differently from caps that have been in the warehouse for a while. This can be even more true with bottles.

There may not be much that can be done to resolve these problems but awareness of the possibility will at least provide a starting point for troubleshooting.

There are thousands of different cap styles and probably as many styles of neck finish. I have addressed some of the mechanical issues of capping above. However, unless the capper, container, neck finish and product are designed or selected to complement each other, capping will always be problematic.

This article originally appeared in the July 2002 issue of Food & Drug Packaging magazine

John R Henry CPP

7 How to solve the top 5 common cartoning problems

The carton is often the key component of the total package from the customer's point of view. It is what the customer sees on the shelf. It conveys information about what is in the package explicitly through its graphics.

Manufacturing can't do much about the general design of the carton--after all, that's up to the marketeers. But a dented, scuffed, torn or otherwise less than perfect carton will do a world of damage to the image marketing is trying to present. Assuring that the carton leaves the packaging line in perfect condition is a job packaging must face every day. Here are a few tips to make your life easier.

1. Materials

It's always important to match the machine with the components in any packaging process and automatic cartoning is no exception. It's not enough to have chipboard (often called "board") that meets all specifications; the consistency of the board is critical as well. Generally, machines can be set up to run a variety of board but when it's characteristics change, the cartoning operation can become problematic.

A number of things can affect the board. Board is like a sponge and will absorb moisture from the air. As it soaks up this moisture, its characteristics will change. It will become less stiff, it can warp, dimensions can change, glued joints can come loose, or glue applied in the cartoner may not adhere properly, among other things. For best results, carton blanks should be stored in a controlled environment with a stable temperature and consistent relative humidity.

Even the source of the material is important. If board is changed from virgin to recycled, this can cause numerous headaches because it will often run differently. It may be harder or easier to open or fold the cartons. The cellulose fibers used to make the board will also vary from region to region of the world so that a board made from U.S. pulp may handle differently from a board made of Swedish pulp. The key here is that board suppliers should not be changed without careful consideration for how the new cartons will run on your existing equipment.

2. Carton manufacturing

Many cartoning problems can be traced to defects at the converter. A biggie is improper scoring. The score lines need to be just the right depth. Too deep and the carton may not have the rigidity needed to form correctly. Too shallow and it may be too stiff to open easily.

Attention must be paid to the amount of glue used in the joint. If there is too much glue, some may ooze out onto the interior of the carton, gluing the carton blank together. When this happens, even if the cartoner can open the carton, some tearing may occur. If not enough glue is used, the joint can come apart during cartoning. Too much or too little glue will lead to machine jams and downtime.

Vivian Woo of Bivans Corp., a builder of cartoning machinery finds that a frequent problem on new projects is incompatibilities between machine and material. She suggests that one way to avoid this, especially on a new project, is to allow the converter and machine builder to communicate directly. The customer may not have a clear and detailed understanding of what the machine requirements are nor of the converter's capabilities. Allowing them to speak directly with each other--while keeping the customer fully informed, of course--can help them work out some of the machine/material compatibility issues.

3. Changeover

Changeover of cartoners from one size to another can be complex with many precise adjustments required. Many times cartoning machines are readjusted for carton length, width and height. Most manufacturers supply their machines with scales or digital indicators to simplify adjustments and make them repeatable. Where these do not exist, they can be added in the field.

Mere existence is not enough, though. Correct setpoints must be determined, documented and used. Failure to do so will result in inaccurate set-ups with damaged products and downtime as the machine is tweaked into the correct setting.

4. Vacuum

Most cartoners rely heavily on vacuum to handle the carton through the cartoning process. This feature must be in tip-top shape at all times. A number of things can go wrong and they need to be caught before they cause problems.

The vacuum generator is the heart of the system. If it is not generating a consistent vacuum, it doesn't matter if the rest of the system is perfect. A common type of vacuum generator uses compressed air and a venturi. These are simple and reliable but can have problems with clogging.

Clogging can occur when the system is constantly sucking up ambient air. In most plants, and especially around the cartoner, dust is in the air. This dust gets sucked into the venturi and accumulates. It is important that vacuum only be energized when needed to pick up a carton. This minimizes the amount of ambient air sucked into the system.

The discharge of the generator will tend to be noisy so a muffler is usually fitted. This muffler is usually made of sintered brass and acts like a filter. Any dust and dirt that gets through the generator will be caught in the muffler. As the muffler begins to get clogged, vacuum performance will deteriorate. Even slight reductions in flow through the muffler can have a significant effect on vacuum. Mufflers need to be replaced frequently.

Check vacuum hoses frequently for leaks or kinks. Check them while the cartoner is running. I once had a machine where the vacuum worked perfectly but the carton kept falling off the gripper. It turned out that someone had cut the vacuum hose slightly shorter. When the arm extended, the now shortened hose kinked and vacuum was lost. A new hose quickly solved the problem.

"Suckers" are the suction cups that grip the carton or product. They are usually made of rubber, urethane or other resilient material. Over time and cycles of operation, they can lose their resiliency and ability to grip. They need to be frequently inspected for small cuts and abrasions that can cause leakage between them and the carton. Suckers are a wear item and need to be replaced periodically, preferably before rather than after they begin to cause problems

They must be large enough to have sufficient grip but not so large that they extend over scores or edges which can cause leakage. They must also be contoured appropriately for the component being handled. If a round bottle is to be gripped, the sucker must be shaped to accommodate it.

5. Product handling

Erecting and closing the carton is only half the battle. Somehow the product must get inside. This can be done manually or automatically but in either case it requires attention to detail.

Product handling needs to be one of the first things discussed in a cartoning project since so much else will depend on it. The machine builder must get realistic product samples showing the way the product will come to the machine.

For example, a dry product will sometimes be duplicated by filling a bottle with water to match the weight. If the bottle is always vertical, this may work. If it gets tipped horizontal for loading, it will handle completely differently, leading to unpleasant surprises when the actual product is run.

Even temperature of the product can be a factor. If a product is packaged cold, condensation may form on the surface, causing handling problems. The key is to give the machine builder samples that truly represent the range of typical conditions rather than all "good" products.

Careful matching of the machine, carton and product--along with attention to detail in set-up and careful maintenance--will go a long way towards turning a potential nightmare into a dream.

This article originally appeared in the September 2002 issue of Food & Drug Packaging magazine

John R Henry CPP

8 Making your Case with Corrugated

Corrugated cases are the workhorse of the packaging industry. They are the final package for many products and are generally the last stage before the product goes on a pallet.

Corrugated is the "plain Jane" of packaging materials and perhaps not as much attention is paid to it as should be. Strictly speaking, case packing is simply the process of getting the product into the case. In its broader sense, case packing actually combines three operations: forming, packing and sealing. In some situations additional operations will take place, such as partition or slip sheet insertion. This article will use the term "case packing" generically to cover the total process.

There are four major approaches to case packing: side or end loading, wraparound packing, vertical loading (often called "drop packing") and robotic loading.

Side load packers often form and load the case in a more or less simultaneous operation. In a wraparound style case packer, the case is formed around the product and all three operations take place together. Robotic packers use a robotic arm to pick up one or more products, orient them if necessary and place them in the case. When drop or robotic packing is employed, there are often three separate machines, sometimes from three manufacturers: the case former, the packer and the sealer. All four styles have their advantages and disadvantages. All will work well when correctly selected, operated and maintained. None will give satisfactory performance if not.

Six areas to watch for in case packing include:

Machinery Matters

1. Glue

Cases are often formed using hot-melt glue. Cold glue is still used but is less common today. The most critical parameter of hot melt is, obviously, the temperature. Hot melt sets up as it cools. If it is not hot enough when applied, it will not stick as it will already be partially set up. If too hot, it may not set up in the limited time that the two mating surfaces are pressed together. This can be especially critical in high-speed machines where the contact time is limited.

Glue temperature is initially controlled at the glue tank. The glue is then pumped through hoses to the point of dispensing. These hoses will usually have heat tracing to maintain temperature. The entire temperature control system must be properly calibrated and set to assure optimum temperature at the applicator nozzle.

Charring can occur when glue is maintained at too high a temperature, especially with exposure to air. It is avoided by keeping temperatures within recommended limits and avoiding prolonged holding times for the glue.

Glues must also be carefully chosen for the application. There are many types of glue on the market each with different characteristics. Some factors to be considered include substrate type (coated or uncoated), product storage temperature and line speeds. A glue used for a standard case to be stored at normal room temperatures may be unsuitable for a coated case for frozen food.

Glue patterns must be carefully controlled. The simplest way to control the amount of glue applied is by time. This works fine if the velocity of the case is constant. It does not work so well if the velocity changes. Several manufacturers offer systems that will detect changes in speed and automatically adjust dispensing to compensate.

Glue dispensing and cut-off must also be carefully controlled. If the glue, when dispensed, does not cut off cleanly from the nozzle a phenomena known as "angel hair" will take place. These are fine strands of glue that will eventually end up in the operating parts of the machine causing problems in operation in extreme cases or an additional cleaning operation in mild cases.

2. Materials

Corrugated board is subject to many problems in the converting, shipping and storage process that will cause problems in manufacturing. Some of the more common problems include moisture absorption and warping. These problems can be controlled with proper storage.

Another problem that occurs in the converting process is quality control. Occasionally a purchasing department will be of the opinion that "it's only corrugated" and look for the lowest-cost supplier. This can result in case blanks with variation in size, improper scoring, cases improperly glued or glued together and other problems. This is a case of "penny wise and pound foolish" since the savings from cheaper materials will be quickly eaten up by excessive down time. Automated processes require automated grade materials. Anything less will cause problems.

3. Servo vs. mechanical systems

In the past, a single motor drove most machinery with levers, cams, chains and other methods controlling all the machine motions. Alternately, some functions were actuated pneumatically. These systems are rugged and reliable but may not incorporate the degree of flexibility required in these days of mass customization and frequent changeover.

Today, more and more manufacturers are replacing mechanical drives with servo motors. Servos have many advantages in terms of simplicity and flexibility. This results in less downtime for maintenance and repair as well as greater ease in changing between products or to new products. Mechanical adjustments are replaced by programming. Another advantage is that some machinery manufacturers incorporate an Internet connection into the machine. When problems occur, their technical service staff can connect directly to diagnose and often correct problems.

The downside to greater sophistication with servo drives is that the plant personnel may not have the knowledge and skills required to maintain them. When this occurs, savings from simplicity may be consumed by the need to bring an outside service technician.

4. Lane dividers

Packaging lines typically handle products in a single lane while case packing typically requires that the product be divided into multiple lanes. Lane dividers can sometimes be temperamental both to set up and to operate.

One common style of laner uses a swinging arm, which distributes the product to the various lanes as required. This arm must line up exactly with the lane for smooth transfer of the product.

One solution I have seen uses a proximity sensor on the end of the swinging arm. A bar with shaft collars is placed in a fixture above the lanes. The proximity sensor senses the shaft collars, which correspond to each lane and stops accordingly A bar/collar set is made up for each size and, to change sizes, the bar is simply swapped out.

The lane width is usually set using individual spacer blocks. This leads to a multiplicity of parts that can easily be avoided by cutting the spacers out of a single piece of plastic. The block is simply slotted appropriately for the lane width. This avoids problems of trying to locate a number of loose parts, as well as assuring that spacers from different sets are not mixed. This can be carried a step further by making two sets of spacers on the same block. Changeover is then accomplished by simply flipping it over.

Product backpressure in the laner section must be carefully controlled, especially in drop packers. Typically the packer grid is not powered and relies on the backpressure to push the containers into position. If there is insufficient pressure, containers will not be positioned correctly and when the grid opens, they will jam. Conversely, if there is too much back pressure, containers may jam in the grid and not fall when it opens. Backpressure is largely a function of conveyor speed, lubrication and chain/container characteristics. When problems are encountered ill drop packers, this is one of the first areas to check.

5. Partitions

Some products require internal case partitions to provide additional protection. This often necessitates a separate machine or hand operation for forming and insertion with all the attendant costs. In many cases a partition can be designed as an integral part of the case and made by the converter. In the packaging line the case is then handled normally on existing equipment. There will be an additional cost for the case blanks but this will often be less than the cost of the case and partition combined and operating costs will almost certainly be reduced.

6. Stacking

Many case packing operations require products to be stacked two or more high. Stackers may be the bottleneck to the casing process. In one plant, they found that the stacker was not necessary. They had been hand placing cartons on a conveyor leading to a stacker/packer. They were able to significantly increase the speed of the process by having the operators place the cartons two high and eliminating the machine stacking operation.

Although case packing seems unglamorous, it can bottleneck even the most elegant packaging line. It is always important to remember that the effectiveness of a packaging line depends on all of its components functioning at maximum levels.

This article originally appeared in the December 2002 issue of Food & Drug Packaging magazine

9 Focusing on 8 common vision system challenges

Manufacturers want to produce perfect products. Part of this quest for perfection is "quality assurance" or developing processes that cannot produce a defective product. The second part of the quest is "quality control" or catching defective products as early in the production process as possible.

Quality control requires product inspection at various stages and frequently this inspection is done either by sampling or, where 100% inspection is required, using humans to continuously monitor an aspect of the product such as cap presence or the correct date code printing.

However, human vision is notoriously unsuited for this purpose. Over the past 20 years, improvements in technology and cost reductions have brought machine vision technologies to a point where they can be used in almost any process. Machine vision systems are typically camera based and include five major components: camera, lenses, software, lighting and product handling systems. All must be properly matched to provide optimum results. If they aren't, the system may cause more problems than it resolves.

Here are some of the major areas of concern when developing a vision project:

1. Application

Simply--what do you want the system to do? While vision systems may seem almost magical in their capabilities, they do have limitations. Speed, product position, feature to be inspected and resolution are all limitations that must be carefully addressed. In some cases, vision may not even be necessary. If the goal is to simply assure the presence of a cap, for instance, a "dumb" photoelectric sensor can be as effective as a "smart" camera with much greater simplicity of hardware and no software at all. Machine vision is as much an art as a science, and selection of a vendor who has the experience to assure that all the parameters are considered is a must.

2. Lighting

In real estate the saying is "Location, location, location." In vision systems, it is lighting, lighting, lighting. Lighting will make or break a vision system and is key to successful applications. No vision system should be installed without its own dedicated lighting system.

While there may be instances where ambient lighting will work, there is no guarantee that the ambient lighting will remain constant. If maintenance personnel change from Cool White to Daylight fluorescent tubes, the vision system may go from working well to not working at all. The lighting and camera must also be shielded to prevent interference from ambient light. In one plant, the system went haywire every morning at 10:00 and then corrected itself by 11:00. The problem was eventually attributed to direct sunlight through an unshaded window.

With lighting, many factors need to be considered. For instance, should the lighting be direct or diffuse? Direct lighting shines directly on the product and is necessary when the product surface requires inspection. Verification of a printed date code will normally require direct lighting. A common problem with direct lighting is that the light may reflect back into the camera. This can be avoided by placing the lighting at an angle or by mounting the camera at right angles to the product and using a 45-degree mirror.

Diffuse lighting (sometimes called backlighting) illuminates the product from behind and is generally more appropriate when the edges of the product must be inspected. Typical diffuse lighting applications include dimensional measurement or inspecting for flash on molded parts.

Lighting color will also play a role. If there is not enough contrast between the feature to be inspected and the background, a colored light source may help.

Most vision integrators will have access to lighting laboratories where sample products can be tested and the optimum lighting design selected. This testing should be one of the first steps in the development of a vision project.

3. Camera

Even though the camera is at the heart of any vision system, most vendors say that it's one of the simplest components to select. Cameras consist of solid state electronics with no moving parts and there is little difference between cameras of similar specifications from differing manufacturers. They tend to be highly reliable and, if they survive their first 50 hours of use, will last indefinitely under normal use.

4. Lens systems

The image captured by any camera is highly dependent upon the lens/shutter combination and a vision system is no exception. Unlike the typical film camera, a vision system shutter is electronic rather than physical. It operates by turning the camera on momentarily to capture the image. Higher production rates necessitate higher shutter speeds. If the shutter speed is too slow, the image will blur. At very high speeds, above 800 or 900 containers per minute, the camera may be left on constantly and a high-speed strobe light used. As the strobe flashes, the camera "grabs" the image. When the strobe is off, the camera sees nothing due to a lack of light.

Lens aperture, sometimes called "f-stop," is another critical parameter. This is the physical opening through which the image is transmitted from the lens to the camera. A smaller aperture will require either more light on the product or a higher shutter speed for the same results. A benefit to a smaller aperture is that focusing the lens on the product becomes less critical. This greater depth of field is important when there can be some variation in product position.

Barrel distortion can occur when the product to be inspected is too large, relative to the lens diameter. When this happens, straight lines near the edge of the lens field will appear curved. This can cause recognition problems for the image processing software. Using a larger lens or moving the lens further back for greater field of view can eliminate barrel distortion.

Filters can be used with lenses to enhance their capabilities. "Sharp cut" filters are designed to exclude all light except for the particular wavelength of interest. In other words, a filter might permit the camera to see a light blue image but not a red image. This is useful when there may be a lack of contrast or conflicting colors on the product to be inspected.

5. Greyscale or color?

A color vision system identifies and processes color and may be required when there is a need to discriminate different colored objects, such as tablets in a blister. Color imaging does require more processing than greyscale imaging and may not always be necessary. If the normal tablet in a blister is light blue and an abnormal tablet is white, color discrimination may be required. If, however, they are light blue and navy blue, a greyscale system can detect them. Many people may think that color discrimination is a nice feature to have but they would do well to keep it simple.

6. Software

The camera and lens capture the image but this is merely the first step. For the image to be useful, it must be processed using software to determine that the actual image matches the desired image. It must also trigger a reject mechanism if it does not match.

For example, in imprinting a date code, some variation in both code quality and position will occur. The software needs to be able to locate the code before it can examine it. Once it has been located, it needs to assure that variations in quality are within an acceptable range. Finally, it needs to trigger a reject signal if it cannot confirm a good code. Sound tough? It also needs to do all this between 100 and 2,000 times a minute. To achieve these speeds, the software does not verify 100% of the image, but only enough to assure that the feature being inspected is acceptable. The algorithms used must strike the appropriate balance.

7. Operator interface

A key component of any automated system is the operator. Because vision systems are designed by engineers, they sometimes forget about the operator. In selecting a vision system, attention must be paid to the human-machine interface.

In a typical installation, the vendor sends a technician to provide initial training to the client's mechanics and operators.

Vision system control panels must be user friendly and intuitive to understand if they are to be used correctly. Graphic operating panels are preferable to text-based. Well-written and profusely illustrated operating and maintenance manuals are a must.

8. Fail-safe rejection

One final comment applies not just to vision but to any type of inspection system. They should always be designed to fail-safe. That is, they should be designed to accept good product rather than reject bad product. One way to do this is to arrange the system to reject every single product that enters the inspection station. The inspection system then inhibits rejection of all good product. While this may seem pretty basic, it is a point often overlooked. The cost of falsely rejecting a good product is minimal. Acceptance of a bad product, especially for a pharmaceutical manufacturer can literally mean life or death.

Vision systems have progressed significantly over the past 20 years and continue to improve in both price and performance. They are not yet plug-and-play and, if a system is improperly designed, it may actually cause more problems than no system at all.

This article originally appeared in the January 2003 issue of Food & Drug Packaging magazine

John R Henry CPP

10 Solve common tamper-evident sealing problems

Is your product safe from attack?

In the best of circumstances, product tampering will be extremely expensive for a company.

Even in these times of heightened homeland security, the risk of product tampering is higher than ever.

Food, beverage, pharmaceutical and medical products are probably at the greatest risk but manufacturers of other nonconsumable products cannot be complacent either. All packagers need to be aware of the problem and take steps to reduce exposure.

First, realize that there is no such thing as a tamper-proof package. Some packages, such as metal cans or glass bottles, approach this. Other types of packaging can make tampering difficult but none are 100% tamper proof. For most manufacturers, trying to design a fully tamper-proof package will result in frustration and high cost without ever achieving the objective.

The goal should be tamper resistance and, more critically, tamper evidency. Tamper resistance means that it is difficult to compromise the contents of a package. Tamper evidency means that, if the contents are compromised, it is clearly evident to the consumer that tampering has occurred prior to use. The use of a foil seal on a cough syrup bottle does not prevent someone from

Machinery Matters

opening and resealing it. What it does do is show the consumer, by its absence or damage, that someone has been into the bottle.

Various techniques are available to provide tamper resistance/evidency on bottles and jars. This article will discuss three common techniques: induction sealing, glue sealing and external shrink seals. We'll point out some of the production problems that can occur with each--and how to correct them.

Induction sealing

As induction sealing systems have become more compact and simpler, this method has grown in popularity. Induction sealers are simple machines, consisting of an electrical coil mounted over the conveyor. The magnetic field from the coil causes the foil seal in the cap to become hot, melting the coating and bonding the seal to the container lip.

Set-up Parameters:

Successful induction sealing depends on the amount of energy transferred from the coil to the liner. The coil current is easily regulated from the control panel. The amount of energy transferred can be a bit trickier and this is where seal failures generally originate. Energy transfer is determined by the proximity of the cap to the coil and by the amount of time under the coil. In setting up the sealer, it is critical to assure that the clearance between head and cap is consistent between set-ups.

If using a gauge to check clearance, be sure that the bottle being used for set-up is typical. A container that is slightly taller or shorter can result in improper adjustment. A digital indicator to specify the height of the coil relative to the conveyor will generally provide more consistent results and is less subject to operator interpretation. One area that may be overlooked is the need to have the coil parallel to the conveyor. If the clearance is gauged at one end, it may be too high or low at the other.

The third element in successful sealing is dwell time. In most applications it's a function of conveyor speed and it is critical that these conveyors have tachometers to allow precise speed setting.

Cap Torque:

Induction sealing relies on the pressure of the cap forcing the seal into intimate contact with the container lip. If insufficient application torque is applied, the seal will not bond correctly. A related problem occurs if a cap is cocked or if the container lip is deformed or damaged. These are not sealer problems, they are capper or container problems.

Heating:

Induction sealers will heat any metals within their field. This can result in a safety hazard if metal components such as guide rails are placed too close to the sealing coil. If heating is noticed, either move them further from the coil or replace them with plastic components.

Container Condition:

Container conditions will vary. One manufacturer may use a mold release and another not. As plastic bottles age, pigments may migrate to the surface. These and other conditions should be held to a minimum but can probably never be eliminated. Some adjustments may be necessary to machine parameters to compensate.

Loose Caps:

Induction sealing causes caps to loosen as the foil coating melts. In most cases, you'll need to use a cap retorquer after sealing to retighten caps to specification.

Missing Liners:

Missing liners are generally a quality issue and should be discussed with the cap supplier. In some instances the capper itself may cause the liners to fall out during cap orientation. It is easy to tell if this is happening, look for liners on the floor or in the capper. Solutions include more aggressive adhesives to hold the liner in the cap or gentler handling in the capper.

Fortunately, this problem is easy to monitor. After sealing, a capacitance sensor can easily check each bottle and trigger a reject mechanism if the foil is missing. These detectors are available from most induction sealer manufacturers.

Glue sealing

Glue sealing also applies a seal under the cap. Instead of heat, it relies on cold glue for adhesion. It has the advantage of being simpler than induction sealing, requiring no heating coils or retorquing. Its disadvantage is that it is messier and requires clean up.

Set-up:

Glue applicators need to be properly maintained and adjusted to assure that the application roller has the correct amount of glue. Too much and it will drip, too little and the seal will not adhere. The key element is the doctor blade, which controls the thickness of the glue. If it is not parallel to the glue roll or nicked, it will cause uneven application to the roll and ultimately to the container.

Clean Up:

While proper adjustment will reduce dripping, it never completely eliminates it. When drips occur, they will travel throughout the entire line. Routine removal and cleaning of the conveyor chain may be required to prevent glue buildup. An alternate solution is to mount the glue applicator on a 2- or 3- foot section of belt conveyor mounted alongside the main conveyor. Containers transfer onto it, receive the glue and transfer back to the main conveyor. Spills or drips are isolated on an easily cleaned belt. Mounting the system on castors allows it to be removed to a washroom for easy clean up.

Torque:

Glue sealing also depends on the pressure of the cap to press the seal to the container lip. Loose or cocked caps, deformed lips or similar problems will result in less than perfect seals.

External shrink seals

Internal seals have the commonly cited disadvantage of not being visible to the customer until they have opened the product. External seals provide a visible assurance to the customer that the product has not been opened.

Shrink bands, as their name implies, are placed over the cap and the neck of the container and shrink into place holding them together. Most people think plastic when they think of shrink bands and this is by far the most common material. An alternative is cellulose bands, which are applied wet and shrink as they dry. They do not require heat, which is a significant advantage when banding an explosive or highly flammable product such as oxygen bottles.

Potential problems to watch for with heat shrink bands include:

Container Design:

The most common problem with bands is improperly designed components. If the container has a straight neck with no lip, the band can be removed, intact, with the cap. The undamaged band can then be replaced. When a neck band is to be used, the container should have a lip for the band to cling to. A straight-sided cap can cause a similar problem if the band does not come up and over the top.

Improper Shrinkage:

For effective banding, the right amount of heat must be applied in the right location. Too little heat and the band will not shrink securely to the container. Too much and it will break along the perf lines. If the heat is not evenly applied, distortion of the band can result with one side shrinking more than the other. This is especially critical when the band has graphics.

As the band shrinks on some containers it may have a tendency to ride up over the lip.

A jet of hot air directed at the lower edge of the band will preshrink it around the lip holding the band in place in the tunnel.

Perforations:

To provide tamper evidency, the band must be destroyed on removal. Perforations ha the band serve to assure that the band is destroyed but they also make it easier for the customer to open. Perforations must be weak enough to break on opening while strong enough to maintain their integrity during application and storage.

Storage:

It doesn't take much heat to begin the shrinking process. Sometimes simply storing the band material in a hot warehouse may be enough. When problems are found with band material, check the storage conditions.

Many options exist for providing tamper evidency on bottles and jars and we have only touched on a few. What is not optional today is that some form of protection must be provided. The type of protection chosen will depend on many factors. In all cases, the protection, container, cap and manufacturing process must work together as a system to assure optimal results.

This article originally appeared in the March 2003 issue of Food & Drug Packaging magazine

John R Henry CPP

11 How to augment, troubleshoot your timing screws

Packaging machines cannot act on a product until the machine has it under control. Many times problems with packaging operations are not with the machine but with the way the product is handled. A labeler may consistently dispense a label with 1/64-inch accuracy but if the product position varies during placement, the label position will vary as well.

Timing screws are one tool that can be used to assure accurate positioning.

Timing screws are just that--screws consisting of a helically threaded cylinder. The product to be handled rides in the groove or pocket formed between the threads. Normal fastening screws have a constant pitch or separation between threads. Timing screws will often vary the pitch for container separation or other reasons.

A screw may start with almost no separation between pockets at the infeed and spread out to 6 inches more at the discharge. Timing screws can perform a multitude of other tasks as well, including combining, diverting, pausing, inverting or collating products. Screws routinely handle bottles and cans in a variety of sizes and shapes but can also handle many other products. One unusual application used a pair of screws to carry broccoli stalks past a cutter.

Terminology

Timing screws are most frequently made of plastic such as Delrin or polyethylene. The advantages of plastic are that it is rugged yet easy to machine, available in a range of colors and relatively gentle with the product being fed. Steel cans running at or high speeds will quickly wear out a plastic screw and stainless steel may be a better choice here.

Other materials including aluminum and wood, which can be used where appropriate. Screws may be directly driven by the machine they serve or driven independently and synchronized electronically. They may run continuously or may start and stop to position or release containers on demand.

There are four important terms to know about timing screws:

> **Outside diameter (OD)**--The maximum diameter of the screw, normally measured at the discharge end.
>
> **Root diameter (RD)**--The diameter at the start of the thread or the minimum diameter of the screw.
>
> **Pocket**--The space formed between the outside and root diameters.
>
> **Pitch**--The distance between the centerline of the pockets. This is the most important dimension as it determines the spacing of the containers to the next operation. Unless otherwise specified, it refers to the pitch at the discharge as it may be different in other parts of the screw. Discharge pitch is measured from the end of the last thread to the leading thread of the next pocket. Otherwise pitch is measured from the leading edge of one pocket to the leading edge of the next pocket.

How they work

The screw's most common use is spacing and synchronization. A typical application is a capping machine that requires incoming bottles to be spaced 6 inches apart and placed into matching pockets on a starwheel for transfer to the capping carousel. Bottles come to the screw infeed with random or no separation and the screw needs to make sense of this.

The infeed of the screw is the most critical section and requires careful attention to design. Once the bottle is in the pocket it is generally fairly easy to control but getting it into the pocket may require some magic.

Usually it is best to run a screw with some backpressure at the infeed. You can do this by placing a sensor upstream from the screw to stop the capper if there is an insufficient quantity of bottles. Proper positioning of this sensor is also important. If it's too close to the infeed, it will cause the capper to start and frequently as the backlog is worked off. If it's too far upstream, it might not leave sufficient space for the filler to discharge.

The balance of the machine speeds will play an important role but they can seldom be matched exactly unless they have a common drive. In that case, a transfer screw can be used to take bottles from the filler and convey them between machines on a one-to-one basis.

If the container is round and speeds are not too high, the screw can be designed with a random infeed. This type of infeed starts with the root diameter and no threads at all. As the container enters, it gradually comes under control of the threads, which then deepen until the pocket is fully formed. One of the reasons this infeed works is because of the gap formed between bottles by their circumference.

Square or rectangular containers, especially cartons, have a different problem. By their nature, there is little or no gap for the thread to enter. One design that overcomes this problem is the inverse taper or cone infeed. This forms a cone with the large diameter first then tapering to the root diameter. As the carton exits the cone section, it is twisted forward opening a gap between it and the next carton. This gap then allows the screw thread to get between them and gain control.

One of the most interesting things about timing screws is their ability to simply perform many otherwise complicated tasks. One use of screws is to stop bottles for intermittent operations such as filling, cotton insertion, neck banding and check weighing. You can do this by using a clutch-brake mechanism to start and stop the screw.

Where feasible, a better way is to use a dwell screw. In a dwell screw, the screw is cut with a normal pitch until about the center of the screw. At this point, the forward pitch of the screw changes to zero for about half a revolution. While the screw continues to rotate, the container stops moving.

Expanding their use

Here are three ways timing screws augment container control:

Machinery Matters

1. Collation

Screws can be used for collating containers into a cartoner or bundler. The containers might be the same, in which case a single screw with a double or larger pocket is used to control two or more containers at a time. They can also be used to collate different containers, such as a two-part product. A pair of screws, synchronously driven, accepts containers and discharges them together.

2. Laydown

Horizontal cartoners normally require bottles to be laying on their sides for loading. It can be difficult to lay bottles over, especially while keeping them in synchronization but timing screws offer one simple way to do this. The screw is designed normally except that at the last pocket, the root diameter increases rapidly and the pocket disappears, laying the bottle over.

3. Mass feeding

Some applications require taking containers from massflow conditions, such as the discharge of a vial sterilizer, and converting them to a single file. Timing screws mounted at a right angle to the mass of vials work well at this. The vials push into the pockets and are moved to the side where they exit the screw single file onto a conveyor.

Troubleshooting tips

If timing screws aren't performing at their peak, here are four ways to troubleshoot their operations:

1. Bottle tipping

Perhaps the most common problem encountered with timing screws is containers tipping in the pockets.

This is often caused by not understanding how screws work. A common misconception on the plant floor is that screws should push the container along the conveyor. While there are some exceptions, such as moving a container across a transfer plate, this is incorrect. Normally, the conveyor causes the movement of the container and the screw is used to hold it back. The rule of thumb is that conveyor speed should be approximately 10% faster than the screw speed so that the container is always pushed against the leading edge of the pocket.

2. Acceleration

Acceleration is the change in container speed as the screw pitch increases. It is a factor of the discharge pitch and overall screw length. For a given discharge pitch, the shorter the screw is, the more extreme the angle of the threads. The more extreme the thread angle, the greater the tendency to tip the container.

3. Screw rotation

Screws must be designed and installed with the correct rotation. The screw must rotate down on the container. If it does not, it will tend to lift the container and the angle of the pocket will cause it to tip.

4. Pocket design

Screws need to mimic the container shape to closely confine it in the pocket. As the pocket is cut on an angle and the container is vertical, this is often an approximation. Careful attention must be paid to pocket design to accommodate the two and assure smooth operation.

Timing screws are often taken for granted and this is unfair. They are the workhorse of packaging permitting many operations that might be difficult to do in any other way. A screw is simply a rotating piece of plastic with no mechanical components other than the drive. It is hard to imagine anything simpler. Given their manifold capabilities, it is also hard to imagine anything more useful.

This article originally appeared in the June 2003 issue of Food & Drug Packaging magazine

12 Powder filling problems – Solved!

Filling dry products presents special problems. These problems include flowability, density variations, lumpiness, dusting and more. This article will look at some of the more common problems and ways to control them. For the purposes of this article, we will look at all dry products filled by either weight or volume as opposed to actual count. These products can range from fine powders such as flour or talc, through granular products such as soap detergents to relatively large products such as candies.

There are two broad general classifications of dry fillers. Weight based fillers determine the product's net weight either before or after it is in the container and control it to meet specification. Volumetric fillers determine the product's physical volume, generally prior to filling. One technique uses a cup of known dimension, filling it and then dumping into the container. Another volumetric technique uses an augur or screw. The screw displaces a certain amount of product per revolution and by precisely controlling the amount of rotation, a specific amount of product can be dispensed.

Weight and volumetric techniques are sometimes combined. For example, an augur may have a dual control. A container is positioned on a weigh cell under the filler and tared. The augur dispenses approximately 90-95% of the desired fill based on revolutions. At this point, the scale takes control of the filler and runs the augur until the desired net weight has been achieved. This allows the speed of volumetric filling to be combined with the accuracy of net weight filling.

So what are some of the problems to look out for?

Product Density

Many dry products are sold by weight but filled by volume. This is fine as long as product density remains constant but it often doesn't. Some products will compact in the filler, increasing density, resulting in an overfill. Others may aerate or "fluff", reducing density and resulting in an underfill.

Density control begins in the transport of the product to the filler. Screw conveyors and pneumatic conveyors can be convenient means of transport but they can also aerate the product. If they are used, they should be used to bring the product to a vibratory pre-feeder which will de-aerate the product. When using an augur filler, the augur itself can cause aeration. To minimize this, the height of the augur hopper can be extended to provide more compaction. Where too much compaction is a problem, some augur fillers use agitator blades in addition to the augur to keep the product loose.

Fill tube or "cup" style volumetric feeders require the product to flow into a fill tube of a certain volume. Scraper blades and/or vibrators may be used to assist the product into the fill tube and to maintain a uniform density.

One way to get around the density problem is to use a weight based filler. However, even with a weight filler, problems may occur when a product becomes aerated and will not fit into the container, spilling out the top.

Level control

The level of product in the filler hopper will affect the density and flow. If too high a level is maintained, excessive compaction can occur. Too low and the product may not have enough backpressure to flow. There are several ways to measure and control powder levels but the simplest is a capacitance sensor. When the level rises to touch the sensor, it senses it and stops the infeed, usually after an adjustable time delay. The time delay is used to prevent the feeder from excessive cycling. A dual system, with a high level control to stop the infeed and a low level control to start it will provide more precise control. A third, or "low-low" sensor can be used to stop the filler and activate an alarm when there is insufficient product in the hopper.

Regardless of the method of control, the feeding mechanism should be adjustable to allow control of the feed rate. This rate should be set slightly faster than the filling rate to maintain the hopper full. If too fast, it can result in excessive swings in product level. If too slow, it may not be able to keep up with the fill rate. Always remember that the the filler can operate to a steady state, the better it's performance will be.

Flowability

Some products will flow smoothly while others will not. The rule of thumb for whether a product is free-flowing or not is to compress a handful like a snowball. If it will form a ball, it is not free-flowing. Non free-flowing products present some special difficulties in handling. Care must be taken to avoid restrictions where the product can build up or clog. Increasing the angle of hopper walls to reduce product compression can help. It may be also helpful to use special coatings such as Teflon to reduce sticking.

Products can rat-hole or bridge in hoppers. Rat-holing occurs when product around the periphery of the hopper does not flow and a hole develops in the center of the mass. Bridging occurs when the mass of product in the hopper gets compacted to the point that it will not flow at all. Vibrators mounted on the hopper can help control these effects.

A key to accurate filling is to maintain consistency of the product. A product with a tendency to form lumps will not flow smoothly. A container will be almost full and then a large lump may come through causing an overfill. Screens, vibrators, agitators and other techniques can be used to break up these lumps.

Product deterioration

In any filler it is important that the product not be degraded in handling. One common problem with granular products is the breakage of the particles causing "fines" or particles of smaller than desired size. Care should be taken to handle the product as little as possible as well as handling it as gently as possible to minimize fines. For example, agitators and vibrators should not be operated except as necessary. Under the best of circumstances, fines can still occur. When they cannot be eliminated, they should be removed by passing the product over a screen of the appropriate mesh to allow them to fall out.

Dust control

With powder products dust can be a major problem. Flammable dust will create a severe explosion hazard and explosion proof equipment construction is recommended. Abrasive products, such as a scouring powder, can get into machine components and cause excessive wear. If nothing else, dust will be a nuisance. Some fillers will measure the product externally and then drop it into the container in a single dump. When this happens, the product entering displaces the air already in the container. As the air tries to escape upwards, it will entrain some of the product trying to flow downwards. Clouds of dust will be the invariable result. In this situation the only solution is a dust collector with properly designed collectors to catch the dust as it escapes.

The best way to control dust is to not generate it in the first place. Bottom up filling puts the filler discharge at the bottom of the container raising it as filling takes place. Alternately, the discharge may remain stationary and the container is first raised then lowered during filling.

Moisture content

Powder products tend to be very hygroscopic, absorbing moisture. Moisture levels will affect density as well as flowability. Products need to be manufactured, stored and transported in controlled humidity environments. Controlled does not necessarily mean low humidity. It does mean constant humidity levels to avoid changes to the product. One thing to keep in mind is that there may be dramatic shifts in humidity levels between summer winter. When it is cold outside, the inside, heated, humidity may drop close to zero. Low humidity and moisture content can be as much of a problem as high. A very dry product in a dry environment will have a tendency to static electricity which can inhibit flow. In extreme cases it can result is fairly large static charges and sparking. Not a good thing, especially if the powder is flammable.

Augur design

Fillers can look fairly simple but there are many subtleties in their design. The augur in an augur filler is not a simple dumb screw but a precision engineered component designed to match product characteristics as well as filling speed and volume requirements. One size does not fit all, even on products that may appear similar. When running any new product, it is always a good idea to send samples to the machine manufacturer for their recommendation about the proper augur.

Cup design in a cup style volumetric filler is likewise important. Some products will need a relatively shallow, large diameter cup for best results. Other products, may require a deeper and smaller diameter cup for best results. Again, it is a good idea to have the machine manufacturer evaluate the product and recommend the appropriate cups.

Document running parameters

Most fillers have a number of parameters such as speed, hopper level, agitator speed and others that must all interact to provide the correct fill. There will always be one combination of settings that will result in optimal performance. In many cases these are discovered after much trial and error. It is important to keep records of the machine settings each day. A logbook should be kept recording all the possible variables in the setup. These should not only include the initial settings but also the settings at the end of the run to reflect any changes that needed to be made. Consistent settings will give consistent performance

Conclusion

Filling dry products is not really as hard as it may look, though there are other packaging operations that are easier. Careful attention to matching the product, container and machinery initially, along with attention to the details in operation, will result in a successful operation.

This article originally appeared in the September 2003 issue of Food & Drug Packaging magazine

John R Henry CPP

13 Putting the lid on 4 common cap sorting problems:

Cap sorters--sometimes called feeders or orienters--are often taken for granted. Yet they can cause packaging line slowdowns or shutdowns. This article will look at some common types of sorters, typical problems and how to solve them.

Sorters used on cappers can be divided into three basic types: vibratory, centrifugal and vertical wheel.

> Vibratory sorters consist of a stainless or aluminum bowl mounted on springs. Generally they are driven with electromagnets but can also be driven pneumatically for use in hazardous environments.
>
> Centrifugal sorters are similar to vibratory sorters because they are round and horizontal but that's where the similarity ends. The centrifugal sorter's bowl is rotated by a motor and has a rim several inches above the bottom of the bowl.
>
> Vertical wheel sorters consist of a vertical rotating disk with a hopper mounted on a slight angle from the vertical.

Here are four typical problems that can occur with cap sorters and how they can be solved.

1. Vibratory challenges

Machinery Matters

A vibratory sorter depends on friction between the cap and the track. When the track gets slick, the cap will slip and feed rates will slow. Many plastic cap manufacturers will use a silicone mold release in their molding process to improve efficiency. Over a period of time, usually months or years, this mold release will build up in the feeder. It may not be visible but the track will feel smooth and slick to the touch. The release gets into the surface of the track and is tough to remove with any kind of solvent or cleaner.

The best way to restore performance is to use a 150 to 200 grit emery cloth to abrade the surface of the track and roughen it slightly. The sanding action should be across the track rather than along it's length to help assure the best grip. Another cause of slickness is that the caps themselves, especially if made of metal or hard plastic, will polish the track over time. Again, the solution is the use of emery cloth to break the glaze on the running surface.

Some other problems include:

> **Broken springs.** Springs are hardened steel and can fatigue or crack over time. Several springs are usually mated together which can make it difficult to see problems. Remove all springs and carefully inspect them, replacing any that are damaged. Hint: After removing the spring, cracks may still be hard to see. Rapping them with a hammer and listening for a dull sound indicating a crack can help.
>
> **Loose components.** In the event that any components of a vibratory feeder system are not completely tight, vibratory forces will be lost. Check all bolts and tighten as necessary. Generally speaking, it is difficult to over tighten but be sure to use high tensile strength bolts if any need replacements. Check also all tracks and welded components in the feeder to assure that cracking has not occurred.
>
> **Faulty magnets.** Electromagnets are just a coil of wire and generally highly reliable, but they do fail from time to time. To check them, hold a screwdriver or other steel object near the magnet. If no magnetism is felt with the power on, the magnet may be faulty.
>
> **Magnet gap.** If the gap between the magnet and spring is too great, power will be lost. If too small, they will hit very noisily. Generally the gap will be between 0.15 and 0.35 inches depending

on manufacturer and feeder size. Use a feeler gauge to verify that the gap is uniform and in accordance with the manufacturer's specifications. Be sure to remove any rust or dirt from the mating surfaces of the magnet and spring. Even a small amount of rust can cause excess noise and lost performance.

Frequency. Vibratory sorters are tuned to be driven at a specific vibration frequency, generally 120--but sometimes 60--pulses per second (achieved by rectifying normal current) in the U.S. In most sorters speed is controlled by voltage with frequency dependent upon incoming power. In recent years, frequency controllers with accelerometers have become more common. These allow the frequency to be varied while holding the voltage constant.

2. Sorting out centrifugal issues

An advantage of centrifugal sorters is that the cap is exposed and it is easy to check for incorrectly oriented caps in the bowl instead of in the track where they can cause jams. A simple photoelectric sensor can identify caps that are incorrectly oriented. These can then be blown back into the bowl without slowing or stopping the system.

Control of the level of caps in any sorter is important. But it is critical in a centrifugal sorter. Centrifugals depend on the ability to transfer caps up the ramp and onto the bowl rim for orientation. If there are too few caps, they may not be able to push themselves up. If there are too many, they will interfere with each other, causing loss of feed rate and jams. The rule of thumb is that the bottom of the bowl should be covered with a single layer of caps. A pre-feeder or hopper is used to meter caps to the bowl and control the level. It is essential that the paddle switch or other control be correctly adjusted to maintain the level. Speed of the pre-feeder should be set so that it feeds caps slightly faster than they exit the sorter rather than dumping relatively large amounts of caps from time to time.

Air jets are often required in centrifugals to assist in orientation and they must be properly adjusted. This is especially critical with high aspect ratio (height greater than diameter) or irregularly shaped caps. If different caps are to be run, requiring adjustment of the air nozzles, it is critical to have some way of re-establishing the correct air flow. Correct supply pressure alone is not enough, the actual flow must be correct as well.

3. Vertical wheel orienters

Vertical sorters are reliable but there are a couple of things to look out for. The most critical is that the rotating disk and guides are appropriate to the cap. These are generally changeparts, as opposed to adjustment and, if properly identified and if operators are properly trained, this should not be much of a problem.

A critical adjustment is the gap between the disk and the hopper. If this is too large, caps will be allowed to fall through upside down. If too small, caps may get jammed even if in proper orientation.

As in all sorters, speed is important. Too slow and feed rate will not be met. Too much speed will also have the same effect, churning the caps too rapidly to allow them to orient.

4. It's in the cap

No discussion of cap sorting would be complete without a word or two about the caps themselves. It does not matter what type of sorter is used or how perfectly it is set up and maintained. If the caps are not of adequate quality, problems will occur. Some typical problems can be caused by:

> Excessive mold release;
>
> Molding flash;
>
> Loose or broken tamper evidency rings;
>
> Poor dimensional tolerances.

Finally, beware an overly zealous purchasing department. In one case, a company had a capper running a stock cap for years at 75 caps per minute with no problems. Purchasing, without telling anyone, changed suppliers. The new cap met all specifications for dimension, weight and such, but the sorter could not feed more than 50 caps per minute. Nobody could figure out what had happened until someone noticed that the caps were from a different source. Investigation revealed that, while the caps were identical, the distribution of plastic was slightly different causing their balance point, on which the sorter depended, to shift.

The moral of the story is a reminder that automated processes require automated grade components. Even subtle changes can cause fairly dramatic problems. If a sorter that was working correctly one day suddenly begins giving problems for no evident reason, check that the caps themselves have not changed.

This article originally appeared in the December 2003 issue of Food & Drug Packaging magazine

14 Seven ways to smoother bag/pouch packaging

The term flexible packaging is used to describe packaging types mainly constructed from flexible film. This film may be paper, plastic, foil or other materials. Any package rigidity is mainly the result of the materials contained within.

Flexible packaging can include bags, pouches, sachets and sacks. The definitions of the different formats seem flexible, mainly related to size. For simplicity, this article will use the catchall term bag to refer to the various types.

Bags can range in size from milliliters to tons. They will often be manufactured from one or two continuous rolls of material at the point of manufacture. This is generally referred to as form-fill-seal (FFS). They can also be made offline and brought in either as individual bags or in a continuous roll form for automation.

Flexible packaging of all kinds has become increasingly popular in the past 10 to 20 years. Reasons for this include improved materials and machinery, increased consumer acceptance, a need to reduce packaging materials going to landfills (source reduction) and a general and continuing need to reduce overall costs. An example of cost reduction is cereal that is sold directly in a bag, eliminating the carton. Eliminating the carton can also reduce the package size and weight, saving on transportation costs.

Flexible packaging does have some disadvantages, though. A thin film cannot provide as much physical protection to the product as a plastic bottle or a carton can. Bags containing juice or other liquids can be easily punctured and require gentler handling throughout the production process. Because the package is flexible, its shape will be unstable and this can cause difficulties in handling.

This article will look at some of the problems that can occur in bag/pouch packaging operations and how to resolve them.

1. Sealing

Many bag machines use heated jaws to weld the bag shut. The combination of time, temperature and pressure is critical. Too little or too much of any one and proper sealing will not occur.

Too much pressure on a single layer plastic material can result in the seal separating from the bag. In a multilayer bag, the innermost or sealing layer can be squeezed out of the seal. This results in an attempt to seal the secondary layers, a job they were not intended to do. This is even more critical in a multilayer bag as the defect will not be visible and may not show up until the product has been shipped. Some machines have strain gauges to indicate the amount of pressure applied.

Pressure-sensitive films are also available to test sealing pressures. A piece of this film is placed between the jaw and will change color based on the pressure. The color can then be compared to a known standard. Either use a piece of film large enough for the entire jaw or test in several places. Pressure uniformity is as important as the amount of pressure. They do not provide an absolute indication of pressure but can be used to check for repeatability.

Temperatures that are too low will not melt the material sufficiently to get a good bond. Temperatures that are too high can be just as bad. High temperatures with normal pressures can cause the same problem as high sealing pressures. If the material is overheated, it may not cool fast enough and the seal will come apart when pressure is released.

In severe cases, high temperatures can cause the material to char or burn. While digital thermostats used to control temperatures are a must, they are not enough. Testing needs to be done to determine the optimal temperatures and then this data must be noted and used. The ability to achieve an accurate set-point is useless if nobody knows what the setpoint should be.

Dwell time is usually a function of machine cycle speed but is critical to good sealing. Occasionally a machine will be speeded up to increase output with the overall effect that, while bags per minute increase, good bags per minute decrease. This is because the temperature and pressure do not have sufficient time to make a good seal. Increasing temperature and pressure to compensate will sometimes, but not always, help. If machine speed is adjustable, a digital tachometer should be installed to assure that it is running at the optimal speed.

Jaws can be flat and smooth or may have a pattern on them to provide more contact area. In either case, it is important to keep them clean. Buildups of product, material that sloughs off from the bag material and other contaminants can prevent the application of a uniform pressure.

2. Filling

In vertical FFS machines, the bag will be formed with the bottom sealed and the product dropped in from the top. At the time of filling the bottom seal may still be warm and not completely set. If a heavy product--for example, several pounds of nuts--is dropped directly on the seal it can cause it to deform. One solution to this problem is to meter the product more slowly into the bag.

Another may be to angle either the bag or the product flow so that the product does not fall directly onto the seal.

3. Conveying

Conveyors are the bloodstream of any factory, carrying raw materials, work-in-process (WIP) and finished product through the process. Some flexible packaging, such as a small foil pouch, is relatively robust. In most cases, though, their additional delicacy requires that special care be taken when handling flexible packages. It should go without saying that sharp edges need to be avoided. Changes in direction and transfers from conveyor to conveyor need to be carefully considered. Conveyor slippage under the product, while common with bottles and cans, must be avoided at all costs with flexible packages. With flexible packaging, you can actually tear or damage the package if the conveyor slips.

4. Roll handling

If you've ever tried to tear cellophane tape, you may have found it impossible until you make a small nick in the edge. Once the tear is started, it continues easily.

Packaging film acts much the same way. It is especially critical during handling not to nick or otherwise damage the ends of the rolls. A small ding in the end of the roll can show up as a weak spot when the machine is running. As the film runs through its rolls, this weak spot will develop into a tear resulting in a total machine shutdown and rethreading. Handle the film as if it were eggs. Store it on proper racks, use the correct handling equipment and never remove the protective end caps before absolutely necessary.

5. Tracking

Good packages depend on the film tracking correctly through the machine. Generally there will be a series of rollers and guides to control film tension and flow. If any of these rollers are not absolutely straight and square, the film tends to wander off to the side. This can cause loss of registration or improper sealing. One solution that mechanics and operators may attempt is to use guides to push the film back to it's correct position. However, this is almost never a good solution. As noted above, the film edge is delicate and pushing on it can cause it to wrinkle or tear. If the film is not tracking correctly, the only correct way to resolve it is to start at the end of the film, where the bag is formed and work back through each roller making sure each is square.

6. Film tension

Film tension is another critical parameter. If there isn't enough tension, tracking problems may occur. Too little tension can also cause wrinkling. Too much tension, on the other hand, will make it hard to pull the film through the machine causing possible loss of registration or slippage of the pulling mechanism. It also puts excess stress on the film and can lead to tearing or stretching. The adjustment for film tension should be measurable and repeatable. Pneumatic brakes or weighted dancer arms can help with that. Once the optimal tension has been determined for a given product, it should be noted and incorporated into the machine set-up.

7. Pillowing

Some consumers, seeing a fluffy bag of potato chips will complain that the manufacturer is trying to deceive them by making the package look bigger.

Machinery Matters

This is not the case. This pillow effect is used to protect the product. When the package is squeezed, the pressure is on the air in the bag, rather than the product. Some plant personnel may see this pillowing as a defect rather than a protection and try to adjust it out. The key word here is, don't.

Flexible packaging use has grown significantly in recent years. As manufacturers and consumers become more cost conscious and as films and equipment improve, it will probably further replace rigid packaging.

This article originally appeared in the January 2004 issue of Food & Drug Packaging magazine

John R Henry CPP

15 Eight tips for more flexible packaging lines

Here are some ideas -- both new and proven -- on how you can accommodate more variety in your operations.

Prior to the '60s, the Coca-Cola bottling plant manager's job was easy. Coke came in a single flavor and a single 6-ounce glass bottle and line flexibility was not an issue. Today, Coca-Cola comes in more than 140 flavors, containers and sizes. The modern plant manager needs to be flexible as an eel to accommodate them all. Most of you are probably facing similar pressures as the market demands more and different products. Here are eight ideas that can help make your packaging lines more flexible.

1. Changeover

The biggest single cause of downtime in most plants is the time spent changing from one product or package to another. Reducing changeover times brings a number of benefits including increased production and increased capacity. As products proliferate, changeovers become more and more frequent—and thus more critical.

Reducing changeover times can seem a daunting task. In reality, most changeover reduction is a result of many small, simple improvements. Some of these improvements are made to the equipment, such as reducing tool usage. Other improvements are the result of organizational improvements, such as training and documentation.

2. Robots

Machinery Matters

After decades of promises, robots are finally becoming practical for general packaging use. Robots are replacing humans in operations like case packing and palletizing. Robots can often do several tasks at the same time, such as packing product into the case then placing the case on a pallet or taking mixed products from a single conveyor and separating them onto multiple pallets.

"The robot's great flexibility allows them to readily replace human operators on dangerous or dirty jobs," says Ed Goldman, senior vice-president of Foster-Miller Inc. He cautions, however, "There is a tradeoff between flexibility and reliability. The greater the variety of tasks that a robot is expected to perform, the greater the likelihood that it will falter occasionally. End of arm grippers are especially important and must be carefully designed."

3. Servo motors

In the past several years there has been an explosion in the use of servo motors in packaging. Cartoners that used to have one drive motor with mechanical linkages may now have a dozen or more servo motors. Complex speed and motion profiles can be readily programmed for each function. Minor adjustments (for component variance) or major changes (for package size changeover) can now be done quickly and easily from a touchscreen.

4. Inline printing

Many techniques are available which allow packagers to print at the point of manufacture. Thermal, thermal transfer, continuous ink jet, laser and hot stamping are a few in common use. Traditionally, they have been used for variable data such as lot codes. In other applications, they are used for primary product identification.

As speed and quality improve, inline coders/printers are increasingly being used for primary labeling. A hair dye producer required over 80 different labels to cover all the color and country variations. After years of fighting with label inventory, they developed a generic label with no color info. A laser mounted on the labeling machine prints the color info immediately prior to application. Their label stock-keeping units (SKU's) have gone from more than 80 to 1.

5. Pucks

Pucks are plastic cups which hold the container during packaging operations. They are common in the cosmetic and aerosol industries but almost unheard of elsewhere. The great advantage of pucks is that, from the line's point of view, the bottle size and shape does not matter. Additionally, pucks are often less expensive than a set of change parts required to run additional sizes.

6. Pigs for cleaning

Pigs aren't found only on the farm. Pigs are rubber plugs that are forced through process piping squeegeeing the walls along the way. Benefits are improved and faster cleaning, as well as more efficient recovery of residual product from the piping,

Pigs may be the first step in a clean-in-place (CIP) processor. When similar products are being run, a simple pass with a pig may be all the cleaning required. The succeeding product can push the pig through the piping. The pig then serves not only to clean but to separate the two products. Sensors detect when the pig arrives at the end of the line and automatically actuate valving to allow non-stop product flow.

7. Modular machines with docking stations

Modular machine designs allow them to be more readily adapted as products change. Krones' new high-speed rotary labeler is a good example. It uses a wheeled stand to mount the various labeling heads. These stands then dock to the main machine. Changing from a pressure-sensitive to a cold glue label is as simple as swapping out the module. The benefits are that it allows changeover of label type and product to be accomplished externally, while the labeler continues to run.

8 Line layout

Ignacio Muñoz, general director of AutoPak Engineering, talks about the impact of line layout on flexibility in this era of lean manufacturing. "U-shaped packaging lines maximize space usage and create ergonomic advantages because they minimize distances between machines," he says. "Well designed conveyor systems should allow for repeatable accurate positioning of guide rails through mechanical stops or position feedback. Accurate and repeatable control of conveyor speed is also a must."

Flexibility of packaging processes and equipment will continue to become more important each year. Little or nothing can be done about either the shift

Machinery Matters

to higher levels of automation or the explosion in SKU's. Proper consideration of the issues involved and tools available will, hopefully, make it less painful.

This article originally appeared in the April 2005 issue of Food & Drug Packaging magazine

John R Henry CPP

Machinery Matters

Columns

2007-2010

Machinery Matters

16 In or Out?

Hello, and welcome to my new column. This regular feature will look at trends and issues in packaging machinery. My background is in operations and machinery, so the focus will tend to skew that way. However, one issue that I run across frequently is the intersection of package design and operations. I will have more to say about that in future columns.

In the meantime, enjoy. I live on feedback and welcome any comments, column ideas or anything else.

Cleaning is a major issue in manufacturing operations. Failure to clean and sanitize properly results in more than an ugly package. In extreme cases, it can result in consumer illness or even death. In today's plant, two general methods of cleaning are typically used:

1. Clean Out of Place (COP) is a cleaning process where the equipment is disassembled. The parts are transferred to a wash area where they are cleaned and sanitized. They may then be stored or reinstalled. COP systems may be manual or may be automated, using a washing machine.

2. Clean In Place (CIP) requires minimal or no disassembly. CIP techniques include spray balls, pigs and valving systems. Ideally, at the end of the process the operative will press a button and the entire cycle will take place with no further intervention. In actuality, many systems require the connection of piping, sprayballs, drain hoses and the like. Some systems require the removal and manual cleaning of some components.

Machinery Matters

CIP systems have many attractions. Perhaps the greatest attraction is that they remove the operative from the cleaning. This eliminates variability and gives a process that is (or should be) identical, cycle after cycle. CIP also acts as a barrier between the operator and potentially hazardous chemicals in either the product or the cleaning process. Finally, CIP eliminates the lost time as well as machine wear and tear caused by disassembly and reassembly.

CIP is no magic panacea. CIP systems are complex and require a higher level skill set to design and maintain. If the equipment cannot be used during the CIP cycle, it may increase the amount of downtime. COP with multiple sets of parts allows the equipment to run during cleaning. Validation of a CIP system needs to be more rigorous as there is less opportunity for visual inspection of the results. Finally, CIP is not cheap, it will normally incur a substantial capital cost which may or may not be offset by improved operational efficiencies.

So, CIP or COP? It cannot be an automatic decision. The pros and cons of both must be carefully evaluated to determine what is best for each situation.

This column originally appeared in the January 2007 issue of Food & Drug Packaging magazine

John R Henry CPP

17 Power at Your Demand

To appreciate our current level of packaging line automation -- and its as yet untapped potential -- we have to look at how far we've already come.

Prior to the mid 18th century, most work had to be done by manual and animal power. The exceptions such as water and wind power tended to be limited and site specific. It was James Watt's steam engine that gave mankind the ability to generate power anywhere it was needed and brought about the industrial revolution.

The first steam engines were as large as a house. While they could be located anywhere, there was still the problem of transmitting the power from the engine to the machines that would use it. Overhead shafting throughout the plant, with pulleys and belts coming down to each machine, was the solution. This was not only inefficient, it was also noisy, cumbersome, prone to breakage and unsafe.

Electrical motors became widely available in the early years of the 20th century. Initially, they tended to be large and expensive. Overhead lineshafts powered by a single motor could still be seen in the 1950's.

Motors everywhere

As electrical motors became smaller, cheaper and more efficient it was logical to power each machine individually. In packaging, a cartoner would have one motor, which drove all the machine functions via through gears, cams, chains and belts. This works well but these mechanisms all waste energy and cost money to build and maintain.

Machinery Matters

We have seen an explosion of progress in recent years in motors and controls. Servo motors in rotary and linear configurations now perform virtually any required machine function -- and can do so with precise speed, torque and synchronization control.

Their small size allows them to be used throughout the machine. Capping machines can have a servo motor driving each head. This allows precise torque control as well as allowing the torque applied to each cap to be monitored. Cartoners use servo motors for such minor but critical functions as flap tucking. Fillers use servo motors to precisely drive dosing pistons. Servo motors can be used to change machine settings during product changeovers.

Design differently

Servo motors greatly simplify machine design allowing smaller, more flexible machines with even greater functionality.

The machine designer's initial impulse with new technologies is to simply replace the old. That fails to realize the technology's potential. New technologies require new design thinking and creativity must be encouraged.

And don't forget the mechanics and operators. They know mechanical systems. As systems change to electronic, they will need training. Machinery that is beyond their capacity to operate and maintain is a step backward for plant operations.

As the ad used to say, "We've come a long way, baby!" From centralized power to power at the point of use. Smaller, faster, cheaper, more efficient. It's all good -- and continues to get even better.

This column originally appeared in the February 2007 issue of Food & Drug Packaging magazine

John R Henry CPP

18 Machine vs. Materials: From Conflict to Cooperation

In the 30 years that I've worked with automated packaging machinery in a variety of markets, I've witnessed the continuing challenge of getting packaging machinery and packaging materials to work well together.

Sometimes the problems can be subtle. A clear film label was wrinkling as it was applied to a plastic cosmetic bottle. The plant had been running this label/bottle combination for some time with no problem and we could find no explanation for the wrinkling. Painstaking adjustment of the labeler did not fix it. It was almost by accident that we found that the bottle supplier had changed from flame to corona treatment of the bottle. We found an old case of bottles in the warehouse, ran them and found no wrinkling. The slight difference in surface finish was enough to cause problems with the initial tack of the label.

Sometimes the problems can seem obvious. A new spirits bottle was aesthetically pleasing, tapering towards the bottom. Normally there would be no problem running this style of bottle, if the packaging line has the right equipment. Unfortunately, in this case, the line was designed to run a straight sided bottle. These bottles would accumulate, fan and fall over. The first time the plant saw the new bottle was when 50,000 cases arrived on their receiving dock.

In another instance, the package could be run but the design caused operational inefficiencies. A medical packager filled its product in 5 and 10 milliliter vials. If the two vials had been the same diameter, changing between

Machinery Matters

one and the other would have been simple height adjustment of three machines, taking perhaps five to 10 minutes. Because the vials were slightly different diameters, a full three-dimensional changeover was required on multiple machines, taking about an hour and a half.

When I talk with operations people about package design, the reaction I usually get is "Let me tell you about the time...". Everyone has his or her own horror stories. In almost all cases, the problems could have been solved cheaply and easily in the design stage but were costly to fix or live with once they got into the field.

The solution is two-fold:

1. Package designers need to learn what the company's production capabilities are.

2. Talk is cheap; it is the lack of communication that is expensive. Packaging Development needs to involve Manufacturing as early as possible in the design process. Failure to do so is almost a guarantee of problems down the road.

If the package will not run on existing equipment, the plant needs to be able to suggest changes to the package so it will run, or advise what machinery changes (and costs) will be required to keep the new design.

This column originally appeared in the March 2007 issue of Food & Drug Packaging magazine

John R Henry CPP

19 Changeparts versus machine adjustment

A few years ago there was a series of beer ads where friends argued about whether the beer "tastes great!" or was "less filling." The punchline was that both were right.

In a gathering of packaging machine builders or users a similar argument might be heard.

"Machines should be totally adjustable to run any package within a size envelope."

"Changeparts simplify total changeover."

It's not really an argument. There are pros and cons to adjustability, as well as to changeparts (machine components sized for a particular package, such as timing screws, dosing pumps or cap chucks).

Advantages of changeparts include:

> All adjustment is built into the part. They can be quickly mounted and, once in place, no further adjustment or fine tuning is required. Eliminating the tools and skill level required may permit changeover to be done by operators. This frees mechanics for other, more valuable, tasks.
>
> Changeparts can be cleaned and maintained offline while the machine is running. Externalized cleaning means more line uptime.

Machinery Matters

Color-coded changeparts provide a visual indication of proper changeover.

On the other hand:

Each time there is a change in the package, new changeparts must be purchased. This can get expensive fast. The cost of one set of changeparts for a bottling line can be $50,000 or more.

Lead time for changeparts may be as little as two weeks but can be 12 weeks or more.

Changeparts are generally stored offline. Time can be lost fetching them, parts can be lost or damaged and sets of changeparts can get mixed. If not properly designed, changing components can be even more time consuming than adjusting them.

Benefits of adjustable machines include:

New or different package sizes can be accommodated quickly at no additional cost or lead time.

Package dimensions will vary from specifications. A slightly oversized bottle may jam in a dedicated change part. Adjustable systems can better compensate for the variations.

There is never a problem with the wrong parts being mounted on the machine.

On the other hand:

Adjustable machines will generally require a higher level of skill to set up correctly. That is, instead of changing "dumb" parts, the operator must be more experienced to adjust the machine.

So, changeparts or adjustments? No need for argument, both have their place. Most machines will combine some of both. A plant with relatively stable package designs and long lead times might lean towards changeparts. Another plant, such as a contract packager, might need the ability to run different packages on short notice and prefer adjustable machines.

It's not a question of right or wrong; it's simply choosing the approach that makes most sense for you.

This column originally appeared in the May 2007 issue of Food & Drug Packaging magazine

20 Lean changeover: It's as simple as 1-2-3

Packagers today are constantly called on to do more. The days of the "one size fits all" product are long gone as customers constantly demand more product variety. Coke, which used to come in a single version, is now available in 150 or more. This need for product variety is a productivity killer. Changeover between products represents wasted time and wasted profit.

The way out of this dilemma is to move to lean changeover. Lean changeover is changeover from which all wasted materials, labor and time has been eliminated. These combine into wasted profits.

Moving to lean changeover is fairly simple: just eliminate the waste. It's not easy but it is simple. Take a 1, 2, 3 approach.

1. Follow the money

Many different reasons are given for reducing changeover. These can include reduced inventory, improved utilization or better customer response times. While desirable, these are not the primary goal, they are only means to that goal. The primary goal must always be profit maximization. As the movie said "It's all about the Benjamins." Changeover is expensive, usually measured in hundreds of dollars per hour. Even small improvements have tremendous paybacks at the bottom line.

The magnitude of these costs make lean changeover more than a tactic. It is a strategy that separates the world-class companies from the also-rans.

2. Think machine and process

Two paths lead to lean changeover. One is what I call the mechanical path, modifying machines to be more changeover friendly. That is essential but it is not enough. The operational path must be addressed as well. It does no good to reduce the mechanical cleaning and set-up from two hours to one if it still takes two hours for quality to complete the paperwork.

3. Remember: Time is money.

Use a three-step process to reduce changeover time:

First, eliminate all unnecessary tasks. One example of this is to put labels on the leading edge of a shipper case (no adjustment required between sizes) rather than in the center of the case (adjustment required). Standardizing bottle diameters and varying only the height can eliminate much changeover.

Second, externalize everything possible. Externalization means performing changeover tasks while the line is running. An additional set of filling pumps allows cleaning while the line is running. This way, time is not lost during changeover waiting for the pumps to be cleaned.

Third, simplify everything else. Some of this is obvious, such as adding quick mountings and handknobs. Simplification also includes innovative ideas such as multiple photoelectric sensors to make switching between sizes easier.

Edward Deming once said of statistical process control (SPC), "It is not easy, but it works." Moving to lean changeover is not easy, but it does work…one step at a time.

This column originally appeared in the June 2007 issue of Food & Drug Packaging magazine

21 Minimize variation to maximize Quality

In my 30 years as a student, teacher and practitioner of quality, the definition I have usually seen has been "meets specifications," "fit for use" or something similar. A few years ago John McConnell's book on quality, "Safer Than a Known Way," changed my thinking completely.

McConnell teaches a different definition. He says that the absence of variation is the key to good quality. It is possible to have all parameters in specification and still have problems. This is especially the case when there are interactive parameters. The minimum acceptable diameter of a cap could be smaller than the maximum acceptable diameter of the bottle. When this happens, the package will not work. More commonly, the two diameters will vary randomly resulting in some caps fitting more loosely than others.

McConnell agrees that all processes will vary but that the traditional definition encourages a belief, implicit or explicit, that having a process within limits is enough. McConnell's definition emphasizes continuous improvement.

The consumer wants a consistent product that is the same every time they use it. Customers may not consciously know that the definition of quality is absence of variation. They do know that when the product is different it makes them uncomfortable. The only way this can happen is if the machinery and materials used in the packaging process are consistent.

I used the caps and bottles as an example but machinery variability is often an even more critical factor. An obvious factor affecting cap tightness is capper clutch torque but there are other factors as well. Some of these are adjustments, such as speed, chuck height, bottle gripping and alignment.

Some are maintenance related, such as oily chucks or worn bearings. Still others may be due to lack of robustness in machine design.

The variation in any one of these individual parameters may be acceptable. When they are taken together, they may cancel each other, producing a perfect product. Mostly though, they will interact randomly producing a loose cap one moment and an overly tight cap the next. In the meantime, the mechanic is running around, constantly adjusting things to maintain the appropriate tightness. Is it any wonder that so many packaging line mechanics look older than their years?

The key to successful and smooth packaging line operation is to constantly work at reducing all variability, from whatever source. Proper cleaning and changeover including accurate adjustment, proper maintenance and repair and, finally, proper machine selection all go into this. Proper materials must also be supplied.

Zero variability may be an unattainable goal. That does not mean that is should not the goal to strive for.

Good enough never is.

This column originally appeared in the August 2007 issue of Food & Drug Packaging magazine

22 Speed!

The original Mad Max movie showed a garage with a sign that said "Speed is only a question of money. How fast do you want to go?" That thought has stayed with me for 25 years now. It has implications we need to bear in mind when designing packaging lines.

Some people believe that spending more money will result in a faster packaging line. In one sense it will. More money will purchase a 200 parts per minute line versus a 100ppm line. This confuses the goals of the plant. Speed is not really what is wanted. What is, or at least what should be, wanted is a line that puts out more product. While that is related to line speed, a number of other factors also have to be considered:

> Line output is the product of line speed multiplied by efficiency. A 200ppm line running at 50% efficiency will have no more output than a 100ppm line at 100%. This question of efficiency is sometimes overlooked in buying lines. If it is, the line will not produce as expected and money will have been wasted. Machinery purchases must consider the efficiency as well as speed. Efficiency must be considered based on actual field experience with the machine under consideration. This may well be different than the expected efficiency based on manufacturer's specifications.
>
> Line output is also a function of machine interactions and line balance. A packaging line must be considered as a single system, not as a collection of individual machines. Failure to do so will result in less than optimal throughput.

Lines must be designed for two levels of flexibility. In addition to the products the line must run today, tomorrow's products must be considered. In the fast changing marketplace, it is unlikely that today's products will also be tomorrow's. Lines must be designed with this need for flexibility in mind.

Today's marketplace not only brings changing products, it brings more products and line changeover is a significant source of inefficiency. Changeover downtime typically costs thousands or tens of thousands of dollars per hour in lost production and other costs. Additional money spent on rapid, efficient changeover will be paid back in months.

Machinery skill requirements must be matched to the skills of the workforce. Generally, the higher the line speed, the greater the level of sophistication required to operate and maintain it. A plant that is unwilling to invest in the training and equipment to support the line may find its performance disappointing.

Speed really is a question of money but it is not what pays the bills. Productive output is. Packaging machinery must be purchased to maximize output, not speed.

This column originally appeared in the September 2007 issue of Food & Drug Packaging magazine

23 RTM! (Read the Manual)

I never pay attention to instruction manuals and it drives my wife nuts. Using a manual to dope out how to install a dishwasher or set up a DVD player just feels like cheating to me. (I'm also not good at asking directions when lost. It's probably a guy thing.)

This may be OK for household appliances but not for packaging machinery. Unfortunately, too many equipment manufacturers build great machines and then fail to provide equally great operating, maintenance and set-up manuals.

There are several reasons for this:

> Many packaging machines are customized for each application. Sometimes this means modification of a standard machine. Other times it means a customized machine from the ground up. Customized machines require customized manuals which are time consuming to write.
>
> Some machine builders use engineers to write the manual. Engineers are great at many things; technical writing is not always one of them. Other builders will use in-house or outside technical writers who may not completely understand the machine.
>
> It's hard to write a manual before a machine is completed and tested. Once it is completed and tested, the customer usually wants it shipped immediately.

Then there's money. Customers often buy machines on the basis of initial price rather than overall cost. When they don't, builders think they do. Good manuals cost and builders are reluctant to charge for them.

Bad manuals cost even more. These costs accrue over the life of the machine in poor set-ups, improper operation and inadequate maintenance. Unfortunately, the machine cost is visible, the lifetime costs are hidden.

All of these problems are compounded in imported machinery by translations.

It's easy, but wrong, to blame machine builders for these problems. Customers share a lot of the blame by not insisting on good manuals (and being willing to pay for them!). Builders must also do their part by showing the customer the value of a good manual.

A thought on graphics: Machine manuals need to be profusely illustrated with pictures, diagrams, drawings and charts. One picture is worth a thousand words.

Accessibility is key

A good manual, by itself, is not enough. People need to use it. So make machinery manuals easily accessible.

Electronic manuals are one solution. Most machine builders will provide these on new machines. Older manuals can be scanned and saved as PDFs or JPEGS. If on a central server, electronic manuals can be accessed from any computer. Some newer machines allow the manuals to be displayed right on the machine.

Electronic manuals must be protected from uncontrolled editing but the key word is uncontrolled. At some point in their lives, most machines will be modified. Manuals, whether paper or electronic, must be updated to reflect these modifications. Updating controls should include passwords, as well as audit trails of who made the modifications and why.

Finally, there needs to be a mindset of "use the book." The more we know, the more we recognize what we still don't know, so training will help some. It is mostly just a question of instilling the right attitude.

Machinery Matters

There really is a right way to do things. Manuals need to capture this and make sure the knowledge is shared.

This column originally appeared in the October 2007 issue of Food & Drug Packaging magazine

John R Henry CPP

24 Make Machinery Manuals Easily Accessible

My last column discussed the need for good machinery manuals but how do you make sure people use them? If not used, manuals won't do much good.

Accessibility is key. Keeping them in the maintenance shop protects them but it may be more trouble than it is worth to fetch them. Copies need to be kept at the line, preferably on the machine. The easier it is to access them, the more likely it is that they will be referred to.

Electronic manuals are one solution. Most machine builders will provide these on new machines. Older manuals can be scanned and saved as PDFs or JPEGs. Electronic manuals avoid many of the problems of paper manuals, particularly availability. If on a central server, they can be accessed from any computer. Some newer machines have controllers and display panels which allow the manuals to be displayed right on the machine. Network the system to a printer and the operator or mechanic can print the pages needed and take them right to where they are working.

Here's a tip: To be sure that only current copies are used, automatically time and date stamp each page and void them after 24 hours.

Electronic manuals must be protected from uncontrolled editing but the key word is uncontrolled. At some point in their lives, most machines will be modified. Manuals, whether paper or electronic, must be updated to reflect these modifications. Updating controls should include passwords, as well as audit trails of who made the modifications and why.

Machinery Matters

Next question: Can the techs understand the manuals? This is not the question of whether the manuals are well written, which I discussed last column. This is about the qualifications of the techs. Do they have the skills needed to understand a wiring diagram or ladder logic? As machinery and manuals become more complex, higher skill levels will be needed to cope. The only answer to this is training, training and more training. Training needs to be continuous and conceptual (such as, "how servo motors work") as well as specific to a machine. Some plants take the attitude that the techs will pick it up by themselves. A few might. Most won't.

Finally, there needs to be a mindset of "use the book." This may be the toughest issue of all to solve. The more we know, the more we recognize what we still don't know, so training will help some. It is mostly just a question of instilling the right attitude. If anyone has an easy way to do this, please let me know.

There really is a right way to do things. Manuals need to capture this and make sure the knowledge is shared.

This column originally appeared in the December 2007 issue of Food & Drug Packaging magazine

John R Henry CPP

25 New Beginnings

Here are six resolutions that packagers should think about for 2008:

1. Eliminate variation

Quality means the absence of variation (see my column in Food & Drug Packaging, August 2007, p.45) Your customer wants a product that tastes, feels, looks and works the same way every time. When it doesn't they start wondering if it is truly a quality product. Variation occurs in materials, components, product, machine setup and everywhere else through the manufacturing and packaging process. I am experienced enough to know that it will never be completely eliminated, but that has to be the goal. "Good enough" never is.

2. Talk to the packaging floor

Well, not actually the floor but the people on it. They are experts in what they do and too often that expertise goes unrecognized and unutilized. Ask them what they need to do their job better. Get them involved. Not only will you find some great ideas, they will generally be better for being more involved.

3. Eliminate downtime

When you aren't producing product, you aren't producing profit. Sure, you notice the big stoppages. Do you notice the short, 1 to 2 minute "nuisance" stoppages that you have every day? More importantly, do you eliminate them? On a 300-container-per-minute line, five minutes per day of downtime is

360,000 units of lost production at the end of the year. That's a lot more than just a "nuisance."

4. Talk to the package designers

Package designers sometimes come up with great packages that can't be run in the packaging plant. Minor changes at the design stage can avoid later problems when it comes to production. Communication is the key to avoiding this. You need to get involved in the packaging design process up front. After the design has been carved in stone, it is too late.

5. Train

Like the lumberjack who did not sharpen his axe because he didn't have time, too many plants don't take the time to adequately train their people. Training needs to be formal and ongoing. Yes, training is costly. The alternative is worse.

6. Use your noodle

Most importantly, Think! Imagine new ways of doing things. Don't automatically discard an idea because it is not the way it is done in your company or industry. When you run across a good idea, whatever the source, adapt it to your needs and implement it.

So, what are your resolutions for the packaging line? Same old, same old? Or new and improved? The crocodiles are nipping at your heels. It is time to get rid of them.

This column originally appeared in the January 2008 issue of Food & Drug Packaging magazine

26 Sweat the Small Stuff

It is human nature that we are much more likely to respond to a single large event than a number of small events. This is true even though the cumulative effect of the small events may exceed the single large event.

Manufacturing is no different. Most people will react urgently to a machine failure. They will throw people and resources at it until it is up and running again. Small line hiccups are often ignored.

I saw an example of this while doing a workshop. This plant had a number of lines packaging small pouches. While sightseeing in the packaging room, one of the machines ran out its film roll and I had a chance to observe the reloading process. Here is how it went:

After the machine had stopped, it took a minute or two for an operator to notice. He then had to use a dolly to carry a new roll from the other side of the room. Once at the machine, he had to remove the shaft from the old roll and mount the new roll. Finally, he spliced the film and restarted the machine.

Later I asked my workshop group to describe the roll replacement process. What they told me showed that I had seen a typical example. They estimated that it normally took 10 minutes from the time the machine stopped until it restarted.

I asked why they did not have the roll staged, on a shaft, mounted on the machine so that it could immediately be spliced in. I also asked why there was not a yellow stacklight to indicate when there was five minutes left on the roll as well as a red light when the machine stopped.

Machinery Matters

The consensus of the group was that it was fairly minor and that management did not want to spend the money (about $8,000 per machine) to fix it.

Then we did the math:

Take 2 roll changes per shift x 2 shifts per day x 10 minutes per change x 240 days per year = 9,600 minutes per year of lost production. That is 160 hours or 20 shifts per year. Per machine!

So, what do you think? Should they live with these "minor" stoppages? More importantly, what about you? Are you living with "minor" stoppages because they have become just part of the background noise?

My story does have a happy ending. The plant manager visited the workshop and the group explained this to him. He promised to issue a purchase order the following morning to upgrade the machines. He also wanted to know why nobody had explained it this way before.

This column originally appeared in the February 2008 issue of Food & Drug Packaging magazine

John R Henry CPP

27 Evaluate the Cost/Benefit of Training

If you think training is expensive, consider the alternative. One way to classify training is task specific and general. An example of task specific training is teaching operators how to run a particular machine. An example of general training is teaching mechanics effective troubleshooting skills to be used throughout the plant. Both are essential to ensure smooth running lines.

Training, as with anything else, must be justified. It is necessary but management may be reluctant to allocate the required resources. Cost/benefit analysis can be a powerful tool to change their minds.

Costs are usually the easier part of the equation. They include explicit costs such as materials, hiring an instructor or renting a training facility. The opportunity costs of taking associates off the job and/or stopping production may be less obvious. There is also the cost of administering the training and of development if done in-house. Valuing these opportunity costs may be somewhat harder but is not impossible.

The benefits of training are more nebulous. This is where many people may throw up their hands. Take the example of troubleshooting training. This training teaches technicians to correctly identify the problem as well its root causes before attempting to fix it. The first step in estimating the value is defining the expected benefits. These may include:

> Reduction in line downtime per instance. Systematic troubleshooting will usually be quicker than the trial-and-error practices that might otherwise be used.

Reduction in overall line downtime. If root causes are identified and corrected, problems are less likely to recur.

Reduced stress on technicians. A systematic technique, properly applied will reduce stress, reducing mistakes.

Reduced consumption of repair parts. Trial-and-error diagnosis often results in good parts being unnecessarily replaced.

After identifying the benefits, the magnitude of those benefits must be estimated. In a plant where 30 minutes per day are being lost to diagnosing problems, training in effective troubleshooting might be expected to cut that in half. Finance should then be able to translate that 15 minutes per day of additional production into a dollar value. Don't forget: all costs and benefits must be expressed in dollars. It is the only way they can be evaluated rationally.

Comparing the estimated value of the training to the estimated costs then determines whether it is worthwhile.

The final step is evaluating the results of the training. Did it achieve the expected results? If not, the reasons should be determined and used to improve the process for subsequent training evaluations.

Training is not just something nice to do as time and resources allow. It is an imperative in any company that hopes to remain competitive. Cost/benefit analysis is a tool that will help convince management to provide for it.

This column originally appeared in the March 2008 issue of Food & Drug Packaging magazine

John R Henry CPP

28 Metric versus Inch

They say that if we were supposed to measure in metric, King Henry's foot would have been 30.48 centimeters long instead of 12 inches. Unfortunately, we Americans are about the last ones using inch measurements. We don't even use it consistently. Bottles are often described in ounces while that same bottle's cap is almost always sized in millimeters.

For machinery, it can be even more confusing. Many plants have both U.S. (inch) and imported (metric) machinery on the packaging floor. In some cases, a single machine may mix both standards. For example, a U.S. system may incorporate a sub-system that uses metric measurements.

My particular pet peeve is the use of metric tools on inch fasteners and vice versa. For some sizes, they almost work. The metric wrench will be a bit loose but will remove the inch bolt as long as it is not too tight. If it is too tight, it will slip. This can cause a personal hazard such as cracked knuckles. In some cases, it will round the hex corners making it difficult or impossible to remove even with the proper wrench.

Recessed hex setscrews can be particularly difficult to remove once this has happened.

Training will help avoid these problems. One technique is signs on machines identifying them as inch or metric. Proper tools must be provided and clearly identified. Some mechanics will have a tool drawer with a number of Allen style wrenches, in all sizes and states of wear, jumbled together. This makes finding the proper wrench difficult and time consuming. Allen style wrenches are available with a blue coating to clearly distinguish them from the standard

Machinery Matters

black, inch, wrench. These should be purchased for all mechanics and all black metric wrenches collected and discarded. While you are at it, make sure the inch wrenches are in good repair as well.

End, box and socket wrenches can be color coded with paint or tape, blue for metric and red for inch.

Improper fastener use is another challenge. A metric bolt will be misplaced and an inch bolt substituted. It doesn't really fit but the mechanic will attempt to force it. The worst case is when they are able to. The bolt appears to be tight but will fail at the least opportune moment. Otherwise, they will bollix the thread then may drill it out and rethread it in an inch size. This is a guarantee of future problems.

Inch and metric thread gauges should be in every mechanic's toolkit to aid in properly identifying fasteners when in doubt. In the end, it comes down to awareness and training. It's not that hard to keep the inch/metric mix straight. It does require that people pay attention.

This column originally appeared in the May 2008 issue of Food & Drug Packaging magazine

29 Why and how to track OEE

Sharing real-time data with the production team via an easy-to-see display helps keep key performance metrics in mind throughout the shift.

If you work in manufacturing, you've heard the term "Overall Equipment Effectiveness" or its acronym "OEE." Although widely used, I have found that many people don't know what it is. Fewer understand it.

The first thing to know about OEE is that it is not about efficiency. It's about effectiveness. These two terms are sometimes used interchangeably, but are quite different. Efficiency is about doing things right. Effectiveness is about doing the right things. A plant might be efficient at producing tomato soup. But when what the customer wants is clam chowder, the plant is not effective.

Traditionally, plants have focused on efficiency rather than effectiveness. Efficiency, in its simplest form, is the ratio of inputs to output. This is a useful metric but, by itself, does not tell much about how the plant is doing or where the challenges might be hiding. An additional complication is that there are many ways to calculate efficiency. This can make comparisons across different plants difficult.

OEE comes from lean manufacturing toolkit. It combines three key performance metrics to arrive at a single number. OEE is the product of (1) availability, (2) performance and (3) quality.

Availability is the percentage of time spent producing relative to the scheduled available time. A plant running an eight hour shift might schedule

Machinery Matters

one hour for meals and breaks. If downtime from various causes is one hour, availability is six hours production divided by seven hours scheduled—or 86%.

Performance is actual relative to the standard production rate. A standard line speed of 200 parts per minute, over the six operating hours, should produce a total of 72,000 units. If it produces 68,000 units over the shift, its performance metric is 94%. Note that this is total production, not just good production.

Finally, because only good product really matters, quality must be factored in. Quality is the percentage of good production relative to total production. If there are 1,000 rejected products, the quality metric is 68,000 divided by 67,000—or 98%.

When these metrics are multiplied together, they give the Overall Equipment Effectiveness for the line.

OEE = 86% (availability) x 94% (performance) x 98% (quality) = 79%

There are a number of benefits to the OEE metric. The greatest is its simplicity. Anyone can understand how it is calculated and what each of the three components means. It is simple enough that it can be manually calculated. If the data is captured automatically, it can be calculated continuously in real time for display on the line. It is a standardized metric which allows it to be used to compare performance across different processes, plants and even industries.

OEE is useful to give a quick indicator of line performance but is no substitute for analysis. If the line is speeded up at the expense of more rejects, the OEE might look good but few would argue that this is a good situation. Any changes in OEE, positive or negative, must always be evaluated. Identify and maintain positive changes. Find the causes and eliminate negative changes.

So, are you tracking OEE yet?

This column originally appeared in the June2008 issue of Food & Drug Packaging magazine

John R Henry CPP

30 OEE and changeover

My June column discussed Overall Equipment Effectiveness (OEE) and its importance. This month I want to address one of major causes of low OEE.

OEE is the product of equipment availability times performance times product quality. Changeover, the total process of converting a line from one product to another, negatively impacts all three.

Availability is the amount of time the line is available to run production. There are a number of reasons a line may unavailable. Few plants these days have the luxury of dedicated lines. Unless they do, changeover, including cleaning, paperwork, material movement, as well as actual machine set-up, will probably be the biggest single cause of unavailability.

Much of the time spent on changeover is wasted. It is usually possible to reduce changeover times by 50% in six months if a serious program is implemented. One way is to use the ESEE technique. That is, Eliminate unnecessary tasks, Simplify as much as possible, Externalize tasks that can be done while the line is running and perform the changeover Exactly.

Performance is the ratio of the actual to the theoretical production rate. Performance is usually particularly poor immediately after the line is restarted and performance glitches can occur through the entire production run. One cause of poor performance is variability in materials or product. More often, the cause of poor performance is that the line set-up was not performed Exactly. The more exactly the set-up can be performed, the more perfectly the line will run. Yes, "exactly" and "perfectly" don't exist in the real world. That does not make them any less worth striving for.

Machinery Matters

To perform an exact set-up, two tools are needed:

Detailed, written set-up procedures are a must to avoid missed steps and misunderstandings. They must include all set-points and adjustments in quantitative terms. Not "Set the chuck as close as possible to the cap without touching." Rather "Set the chuck 1/16-inch above the cap." The first is subject to interpretation. The second is not.

That's only half the story. In addition to specifying the distance, a means must be provided to measure it. This can be a gauge or other indicator. If not provided, the mechanic is back to guessing. This will cause variability and degraded performance.

Quality in OEE is defined as the number of products rejected for defects. This ties directly to the performance metric. When machines jam, they damage products. When they are not properly adjusted, they will not produce quality product. Some of this may be product that is out of specification, caught and rejected. Some may be product that is marginally within specification but not quite as good as the consumer expects. Worst of all is a product that is out of specification, is not caught and is sent to the consumer.

Today's plant has no choice but to improve OEE. Doing this requires continuous improvement through attention to detail. Changeover frequently offers a tree full of low hanging fruit. To improve OEE, improve changeover.

This column originally appeared in the August 2008 issue of Food & Drug Packaging magazine

John R Henry CPP

31 Ten tips to simplify machine cleaning

All manufacturing facilities should be kept clean at all times for best operation. Food and beverage plants must go further. Cleanliness is not just nice to have. It can literally be a matter of life and death for both customer and company.

Plants must take the extra step to design facilities, machines and processes to assure that they will be cleaned to a level which guarantees uncontaminated product.

Einstein is reputed to have said "Keep it as simple as possible but no simpler." This is a good rule to keep in mind with cleanability. The more complex and difficult the cleaning process, the more likely that failures will occur.

Here are some ideas to simplify the cleaning process:

1. Develop good written Standard Operating Procedures (SOPs) to assure that everyone knows exactly how cleaning is to be done.

2. The smoother the surface, the easier it is to clean. Protruding bolts can be replaced with flathead, countersunk screws. Cracks and holes can be filled with silicone caulking. Horizontal flat surfaces can accumulate dust and liquid. It is much harder to accumulate on surfaces that are slanted or rounded.

3. Cleaning chemicals should be pre-measured to assure their use in proper concentrations. Buckets or sinks used for mixing cleaning solutions should be clearly marked to show how much water should be added.

Machinery Matters

4. Cabinet-style bases on machines may be necessary but can also complicate cleaning. If the cabinet is not sealed, it must be cleaned inside as well as out. A light mounted inside will assure that the cleaner can see what he or she is doing. Dust- and liquid-tight sealing of cabinets can reduce or even eliminate the need to clean inside.

5. Cabinets should either be mounted flush to the floor to eliminate the need to clean under them or raised high enough (6 to 10 inches) to give good access for cleaning.

6. Sanitary conveyors have a closed frame which prevents contamination getting inside. Common in pharmaceutical plants, they should be considered in food and beverage plants.

7. Liquid fillers will always have some spillage. They should be designed so that any spillage is contained. Concave tops with a drain or slanted tops with gutters can help confine spillage.

8. An assortment of hoses, wires and conduits will be hard to clean. Enclosing them all in a single, sealed ductway under the packaging line provides a single, smooth surface that needs to be cleaned. Or use disposable tubing to eliminate the need for cleaning. Another alternative is to use dedicated piping systems that can be easily disassembled. This not only reduces chances of cross contamination, it allows them to be more easily cleaned off-line.

9. Use stainless steel wherever possible. It costs a bit more initially but pays for itself quickly in ease of cleaning and maintenance.

10. Older machinery will often have control panels with an assortment of switches and buttons. More modern equipment will use touchscreens with pictures of the switches. It is easier to clean a single, flat, screen than a variety of controls.

Simplification will not only improve the quality of cleaning, it will reduce the amount of production time lost. Better product with more production? Sounds like a winning combination.

This column originally appeared in the September 2008 issue of Food & Drug Packaging magazine

32 Get Rid of Tools on the Line

I love tools, all kinds of tools. Hand tools and power tools, old tools and new. Out on the line they can make me crazy.

Let's begin with a definition: Tools are implements used to aid in performance of a task. They include things like wrenches and screwdrivers. It also includes gauges and tape measures and even the piece of pipe or the block of wood used as a lever to lift a part.

So what's my beef with tools on the line?

Incorrect tools are used

We've all seen mechanics using pliers on a bolt. Adjustable wrenches are an improvement but are still not right. Use of metric tools on standard parts and vice-versa happens more than most of us would like to admit. Don't even get me started on the tendency of some to replace a full toolbox with a Leatherman style all-in-one tool!

Improper tools can result in personal injury if they slip. It will usually result in undue wear or damage to the part (then in the future the proper tool cannot be used and the cycle continues). Metal shavings generated by pliers can contaminate the product.

Worn tools are not replaced

A worn screwdriver will damage every screw it is used on. Some mechanics will attempt to regrind them but getting it right is impossible. Hundreds or even thousands of dollars of damage can be done with a worn five dollar screwdriver.

Qualifying only certain users: Some plants permit only mechanics to use tools and operators are not permitted to use even the simplest tools. If a bottle jams in a machine guide that requires a wrench to remove, time will be lost waiting for a mechanic. If that bolt can be replaced by a pin or handknob, it may be possible for the operator to clear the jam.

Tools are not available

If tools are not readily accessible, time is lost fetching them. If tools are not kept in an organized toolbox, they may be misplaced. Both situations increase the likelihood that the wrong tool will be used.

What to do?

Eliminate tool usage for common or routine tasks:

Ideally changeovers should be tool-less. Guides and other components that may need to be moved to clear jams should use tool-less fasteners.

Make regular checks to assure that the proper tools, in good condition, are readily available where needed:

Create a simple tool exchange system in place allowing the mechanic to replace worn tools for new with minimal paperwork and hassle.

Don't cheap out with less than top quality tools:

No matter how successful the company, there is never enough money for cheap tools.

Train regularly in proper tool usage and tool maintenance:

The assumption that the mechanic knows how to use tools properly is not always true. Mechanics that once knew may fall into bad habits over time. Supervisors need to be vigilant to make sure tools are used correctly.

The right tools can be invaluable in certain situations but it is best to eliminate the need for them wherever you can.

This column originally appeared in the October 2008 issue of Food & Drug Packaging magazine

33 New versus used

Packagers face the dilemma of selecting new or used machinery. The pros and cons of each can make it a tough decision. Here's a look at some of them.

First of all, the term "used machinery" covers a lot of ground. At one end is brand new, never installed machinery. At the other is machinery that is little more than scrap iron. Then there is everything in between.

The condition of a used machine can be an unknown. How much hidden wear does it have? Are all the electricals up to par? For the knowledgeable buyer these questions can be largely resolved by careful inspection or observing the machine in operation.

Some builders sell "remanufactured" machines. The meaning of remanufactured can vary from company to company but is generally taken to be a step beyond refurbishment with replacement of all wear parts such as shafts and bearings and upgrades to new machine standards.

A number of companies sell used machinery. Some simply buy and resell. Others will refurbish to varying degrees and provide ongoing support after the sale. Establish a relationship.

So, when should you buy used instead of new? There are a couple key factors to consider:

Cost:

Machinery Matters

Used machinery will have a lower price tag than new. This lower initial price may be offset by higher costs of preparation and installation as well as higher maintenance and operation costs. The total cost of the used and new options must be evaluated to get a true picture.

Availability:

Most packaging machinery is built to order. Three-month delivery on pretty much anything is fast. Some machines may take anywhere from 12 to 18 months to deliver. A used machine can usually be delivered in about the time it takes to truck it to the plant. The ability to get online quickly with a new product or with additional capacity is often worth a lot.

There are some drawbacks of used machinery as well:

Technology:

Packaging machine technology is improving daily. New machines use individual servo drives and computer controls for greater flexibility and reliability.

Documentation:

Operation and maintenance manuals along with schematics and parts lists are a must for any machine. These may not be available for older or especially obsolete or orphaned machinery. Some dealers maintain extensive libraries and can supply this. Others do not.

Contamination:

It may be hard to guarantee that a used machine has not been used with a material incompatible with your product. No matter how well it has been cleaned or refurbished, there is always a risk of residual contamination. This may not be important for a cartoner used on a detergent line. It is absolutely critical on a filler to be used with a food product.

Buying used machinery will entail more risk than buying new. But the careful buyer who can evaluate and deal with the risks might find some substantial benefits.

This column originally appeared in the December 2008 issue of Food & Drug Packaging magazine

34 What you see is what they get

We may not have Star Trek's Holodeck yet but we're getting closer. Several technologies shown at Pack Expo 2008 come close to telepresence. Cameras and bandwidth continue to decline in price, making these applications more attractive than ever before. The great thing from a machinery standpoint is that remote assistance no longer needs to rely on verbal descriptions of machine problems for troubleshooting. Now the technician at the vendor can see exactly what the technician on the plant floor is seeing. As the title says, what you see is what they get. An additional benefit is that in many cases this will eliminate the need to dispatch a technician. This will result in getting back up and running more quickly, as well as eliminating travel costs.

PolyPack exhibited a video camera permanently installed on their shrink wrapping system. It is hard to imagine a much simpler system and one wonders why nobody else thought of it. It consists of a video camera installed inside the machine cabinet facing the collation/wrapping area. The camera can be zoomed in and out, tilted and panned remotely. How remotely? Anywhere there is an internet connection.

Another system from Marlen Research Corp. incorporates a video camera, microphone, earphones and laser pointer on a hardhat. The camera feeds to a remote location over the internet via a Panasonic Toughbook PC and Skype, an internet-based telecommunications service. The local technician puts on this magic hat and dials up the remote location. The camera and audio system allow both parties to see, hear and discuss the same thing. A nifty innovation is the laser pointer which is aligned with the camera. This simplifies aiming the camera and eliminates the need to monitor a screen to capture the video of interest.

Machinery Matters

The adage that nothing ever happens while being watched seems especially true with packaging machinery. One can sit and stare for hours but look away for 30 seconds and the problem being watched for will occur.

The FlashBack video system from Hartness Int'l improves on high-speed motion analysis because it eliminates having to scroll through long video segments to find the fault. Although the camera is continuously recording, the system "flashes back" and only saves a video clip of up to four minutes prior to the event and up to 1,140 minutes after it. The entire portable system weighs just 23 pounds and can also be used for training operators and maintenance technicians.

Finally, the coolest application I saw at the show was from Oystar Jones. The technology was pretty standard, it was the application that caught my imagination. The iPod Touch video/MP3 player is designed to play video and sound files. Oystar Jones loaded it with a video showing how to change over a cartoner. Technicians wear the player on their arm and, combined with audio through a wireless headset, are walked step-by-step through the process. This is a great tool for training, as well as daily use. I love seeing standard technology put to unexpected uses. As Einstein said, "Imagination is more important than knowledge."

A picture may be worth a thousand words but a video is worth a thousand pictures.

This column originally appeared in the January 2009 issue of Food & Beverage Packaging magazine

35 Know your costs

Cutting the cost of changeover is something every company can do.

The economy is in the tank and, if they are buying at all, your customers are looking for bargains. If you didn't feel the heat to cut costs before, you certainly should now. Changeover costs can offer a significant opportunity in almost every company.

How much can you save via reduced changeover times? That's a question I ask all the time, but fewer than 15% to 20% of companies can answer it. Companies that do know tell me that it is hardly insignificant. The lowest number I've seen, from a pharmaceutical packager, is $13,500 an hour. A distilled spirits bottler calculates it at more than $32,000 an hour. Other companies in a variety of packaging markets calculate costs in a similar range.

Changeover costs come in two flavors: tangible and intangible. Tangible costs are those which are measurable and whose dollar value can be calculated. Intangible costs may be even more expensive but, by their nature, are difficult or impossible to quantify.

Tangible costs

The biggest tangible cost for most manufacturers will be lost production. If your line runs at 200 products per minute, every minute it is down represents 200 products that you will not sell or earn profits on.

A related cost is lost capacity. If you are selling all you can make, but are losing 10% to 20% of your available time to changeover, reducing changeover provides additional capacity without the expense of new construction, machinery, people and inventory.

Inventory levels are another factor. Some plants will try to minimize the impact of changeover via longer production runs. While this reduces the number of changeovers, it does so at the expense of increasing inventories. At a typical annual carrying cost of 30%, the focus needs to be on reducing, not increasing, inventory.

Better quality changeovers will lead to increased performance. Increased efficiency goes straight to the bottom line.

Intangible costs

The biggest intangible cost is usually reduced responsiveness to your customers. They don't want to hear that you can't satisfy a need this week because you are running something else. The ability to flexibly shift production as needed will help keep a customer from going elsewhere.

A plant stressed by lengthy changeovers, fighting just to get product out the door, can forget that the lack of innovation results in stagnation and decline. Relieving the stress by getting changeover times under control will foster both innovative ideas and the opportunity to try them out. This creates a virtuous cycle where each innovation inspires further innovation.

These examples are illustrative, not comprehensive. Every company will have different models and factors for determining changeover's cost. The point is that there are significant costs to changeover and they must be known. It is only by knowing costs that management can make intelligent decisions. Flying blind on this significant operating cost will lead to bad decisions. In other words:

Know your costs and you will do something about them!

This column originally appeared in the February 2009 issue of Food & Beverage Packaging magazine

36 Don't cheap out

One of life's rules is that you invariably get what you pay for. Some companies let purchasing rule machine selection by focusing on lowest cost suppliers. This policy may cut some visible present costs but it often does so at greater, albeit hidden, future costs.

Anyone who has attended Pack Expo knows that for any packaging task, a range of machinery is available. More than 100 U.S. and even more non-U.S. companies can supply a labeler for a round bottle. Prices range from cheap to expensive. Quality can range from poor to excellent and there tends to be a correlation between the two.

Some companies will try to save money by buying less expensive equipment and sometimes this may work out. Too often, the less expensive machine will also be less efficient. Not much, perhaps. It may only be 1% to 2% less efficient. In a plant which does not closely monitor efficiencies, the difference may not even be noticeable. Noticed or not, this still imposes a cost that can be hundreds or even thousands of dollars per day. Each $100 per day of inefficiency imposes a penalty of $24,000 per year on the plant.

These inefficiency costs will manifest themselves in various ways. They include minor and major line stoppages, longer changeover times, more variability in product quality and damaged or rejected product. Even worse, it may manifest in defective products being passed along to the customer because they were not detected and rejected.

Other costs imposed by the less expensive machine will relate to maintenance and repair. Better machines will tend to have better service and support. In

Machinery Matters

the plant, this means good operating, changeover and maintenance manuals. Some more sophisticated machines will incorporate self-diagnostics that tell the operator or technician what the problem is. Internet-linked diagnostic capabilities can even allow the machine builder to recommend repairs and adjustments remotely. The more expensive machine is also likely have better support in the form of troubleshooting assistance, ready availability of repair and upgrade parts and availability of onsite service technicians.

Training is another way some companies try to save money on machinery purchases. It never fails to amaze me that companies will spend hundreds of thousands of dollars on a complex and sophisticated machine and then assume that a few hours training by the factory tech during start-up is sufficient. It's not. The better machine builders will have comprehensive training programs. They may seem expensive but will quickly pay for themselves.

Machine design is another element often worth paying extra for. A functionally well-designed machine will be easier to operate and repair. An aesthetically well-designed machine will tend to increase pride resulting in better care and better operation. This may be something as basic as purchasing a stainless steel instead of a painted skin. In five years, the painted skin will be scratched and stained. The stainless skin will still look new.

You get what you pay for. In a year the purchase price is forgotten. Value is added every day.

This column originally appeared in the March 2009 issue of Food & Beverage Packaging magazine

37 Get systems and people ready now

I would be the last one to deny that these are tough economic times for many. On the other hand, they do seem to be made worse by the "Woe is me" attitude some take. Why do some people seem delight in feeling bad? There may be challenges but there are opportunities as well for those savvy enough to take them. Don't be like the man who couldn't fix his roof when it was raining and wouldn't fix it when it stopped.

First, realize that this current recession will not last forever. We may even be on our way out of it as you read this. What will you do when your business comes back? Playing catch-up is a losing game. Now, while things are slow, is the time to prepare for the future.

A plant without people is merely a collection of dead machines, yet the first thing to be cut in a slowdown is often the workforce. These people are the heart of the business and will be difficult to replace. It may be hard to keep idle people around so the key is to use them productively.

Some plants do not do as good a job on maintenance as they should. Out on the floor there will be machines operating with temporary repairs sometimes made a year or more earlier. Production demands may have excused the lack of a proper repair then. What is the excuse today with the line at less than full utilization?

Use this slack time to perform thorough repairs and maintenance. This includes cosmetic maintenance like cleaning and painting the machinery. Some of this will be by skilled mechanics but some, especially cleaning, can be done by operators. Don't stop with the machinery, bring the whole manufacturing area up to snuff.

Changeover is less critical at low production volumes. That makes now the best time to work on improving existing practices and developing new ones. Brainstorm and implement better ways. Been thinking of developing a 5S program (Sort, Set in Order, Shine, Standardize and Sustain)? What better time than now. All those standard operating procedures (SOPs), guidelines and checklists that need to be written, this is your chance.

Some plants feel they do not have time for training. Others seem to feel that training can be done once and will last forever. People forget, they fall into bad habits, they don't learn new techniques. Well-trained people are the key to successful operation. Training can be provided by an outside instructor but most companies already have the talent in house. Take your best electrician, for example, and use him or her to teach basics of electricity to the others. There is no better way to learn a subject than to teach it.

None of these suggestions cost much out of pocket. All of them can help justify keeping your skilled people on the payroll.

The company that needs to go out and hire when things pick up is going to be like the race car—sitting on blocks because the tires were sold—when the green flag drops.

This column originally appeared in the June 2009 issue of Food & Beverage Packaging magazine

38 The Snake and the Seer

The snake is flexible and adapts to any shape as necessary. The seer sees into the future. Which kind of company are you?

Flexibility and forecasting negate each other to some degree. A plant with perfect flexibility would have little need for forecasting. The packaging line would be able to produce products one at a time to order. With perfect forecasting, flexibility would be less important as everything would be planned ahead. In the real world of food and beverage packaging neither perfect flexibility nor perfect forecasting will ever be possible.

Forecasting will always be imprecise. The farther into the future the forecast, the more imprecise it will be. Even the best short-term forecast will be disrupted by unpredictable events which will wreak havoc with production schedules. It doesn't matter if the reason is the weather or Wal-Mart, the market demand must be met. If you don't meet it, your competitor will.

Inventory, both raw material and finished goods, is one way to compensate for these disruptions. This does allow unforeseen changes to be met but at a cost. Typically the cost of inventory in the food and beverage industry will be in the range of 30% per year. This covers capital costs, handling, warehouse costs, shrinkage and obsolescence, among other things. That 30% goes on year after year after year. The only way to reduce it significantly is to reduce inventory levels.

Instead of using inventory to remove the effects of uncertainty, reduce the uncertainty by focusing on flexibility. All else being equal, the more flexible company will beat the less flexible company every time. They will do so

Machinery Matters

because they are better able to delight their customers. They will do so because shorter forecasts will be more accurate.

The key to improving flexibility is to reduce the total manufacturing cycle time from customer order to shipment. The first step in reducing cycle time is to measure it and all its components. Some of these components will be productive time—such as entering the production order into the system. Other components will be non-productive—such as waiting for the entered order to begin the next step or picking materials in the warehouse. Line changeover is frequently a big non-productive loss that can usually be reduced significantly.

Put the complete cycle up on a wall either on a whiteboard or a long sheet of paper. Identify times and activities as productive or non-productive. (Some prefer to use "value added" and "non-value added). Then start asking "why?" Why do these non-productive events exist and why do they take so long.

Once identified, begin reducing or eliminating the non-productive times. It will take time and effort. It will involve changing the way things have always been done. It may take some people out of their comfort zone. The end result, greater profitability and growth, will be more than worth the effort.

So get started now and the next time someone calls you a snake, take it as a compliment.

This column originally appeared in the August 2009 issue of Food & Beverage Packaging magazine

John R Henry CPP

39 Lazy manufacturing or lean manufacturing?

"Progress is made by lazy people looking for easier ways to do things."

So begins Robert Heinlein's tale of "The man who was too lazy to fail". The hero, as a boy, looks at his options of working on a farm or going to school. Being lazy, he decides school is easier. On graduation his options are coal mining or college, so off to the Naval Academy. Naval aviation looks like the easiest alternative so he opts for that. Flying a plane is a lot of work so he invents an autopilot and so on.

In most plants, most people work too hard. Some of this involves performing unnecessary tasks; some involves making necessary tasks harder than they should be. Some involves doing normal work in a rush because it had not been properly planned.

This extra work stresses the processes and, more importantly, the people in the plant. This stress leads to mistakes, defective product and reduced overall production.

Most of you are familiar with Lean Manufacturing. Many of you probably have lean manufacturing programs in your companies and if you don't, you must. Lean manufacturing is about eliminating wasted materials, energy, time, effort and work.

The benefits of lean manufacturing brings for the company—reduced costs, increased production and profit—are easy to see. It is sometimes hard for the person on the line to see the benefit to them. They are being asked to move out of their comfort zones and work in new ways. They are thinking, even if they are not saying, "What's in it for me?" Until they understand that, they are never going to get excited about it.

Lazy manufacturing addresses this concern. If it does nothing else, it catches them off guard and gets them listening. More importantly, it encapsulates the entire focus of lean manufacturing. Some people, when they hear lean manufacturing, think that it is about doing more with less or working harder. Lazy manufacturing tells them the opposite. Lazy manufacturing is about working slower, easier and more carefully. Lazy manufacturing is about improving results with less effort. That's what's in it for them.

Lean manufacturing is often implemented top-down and focuses more on big-picture stuff. Lazy manufacturing is bottom-up. Lazy manufacturing is about challenging the folks on the floor, the true experts, to find ways to make their jobs easier. Turn them loose and it is amazing the ideas that they can come up with.

In one plant, operators put small date labels on each carton, peeling each one by hand. An operator suggested a price labeling gun, and this was implemented that same day. In a PET bottle plant, an operator suggested eliminating adjustment of half of the brackets on a thousand feet of conveyor. That was implemented the next day. Operators were not happy about stooping to load cases on pallets. Management ordered a pallet lift that week.

So, lazy manufacturing or lean manufacturing? Of course!

This column originally appeared in the October 2009 issue of Food & Beverage Packaging magazine

John R Henry CPP

40 Rocket line starts aren't rocket science

Are your line starts vertical? Vertical, in this case, refers to the time it takes to go from the line stopped to fully efficient normal production speed.

A truly vertical start goes from stopped to fully efficient speed instantaneously. The fact that this is impossible in the real world should not stop us from striving for it. The ramp-up to full speed has two components: Line charging may require some operator intervention and slow speeds until the line is full.

Once the line is fully charged, ramp-up truly begins. This period is characterized by frequent stops for fine tuning of adjustments, jam clearing, spill cleaning and removal of damaged packages. Some packages that make it all the way through may be rejected as substandard. (Even worse, some substandard packages may not be rejected.) This period can last anywhere from 10 minutes or less to 10 hours or more. In some cases, the run may finish without the line ever getting settled down.

Since the line is producing, albeit inefficiently, some may not be too concerned. "Better to have low production than none at all" seems to be the attitude. I disagree. Ramp-up is expensive, and plants will be better off without it. Direct ramp-up costs include marginal, damaged and rejected product. Indirect costs include less than fully effective operators, and mechanics who are always putting out fires.

Non-vertical start-ups are normally caused by variation in machinery, materials or both. Examples of material variations include varying viscosity of

Machinery Matters

liquids or stiffness of paper and board. Material variations need to be identified, tracked and reduced or eliminated wherever possible.

Machinery related non-verticality comes from several sources:

Some machines require a thorough warmup before they will run properly. If a vertical bagging machine is run before the sealing die temperatures stabilize, problems will occur. One means of addressing this is to activate heaters well before the machine actually needs to run.

Some machine controllers may not be able to maintain the necessary degree of stability. Some machines use 0-10 potentiometers for speed control. Even with a designated setpoint, it is hard to keep the speed just where it needs to be. All moving machines in a packaging line need to have tachometers to precisely indicate speed.

Machines that have not been properly maintained may tend to have excessive play in their movements. This play means that they seldom cycle exactly the same way twice. Worn machines must be taken off-line and refurbished to allow them to operate to tight parameters.

The biggest cause of non-vertical startups in my experience is that the setup was not performed correctly. This requires both standard operating procedures and checklists that show exact settings, as well as proper tools and training to assure that the SOP and checklist were followed.

Verticality, be it after a changeover or simply after an overnight shutdown, is usually relatively simple. Not easy; simple. The principal requirements are good materials, well-maintained machinery and attention to detail.

Getting your line to take off vertically like a rocket is not rocket science.

This column originally appeared in the November 2009 issue of Food & Beverage Packaging magazine

John R Henry CPP

41 Remember: Downtime means downturned pockets

If you had a fountain spewing gold coins, you would do everything possible to keep them flowing at the maximum sustainable rate. Most people don't recognize that this also describes their packaging line. They look at the line and see bottles, cans or boxes coming off the end. In reality, what is coming off the line is money; it only looks like bottles, cans or boxes.

In my experience across many plants and industries, the emphasis is typically on product produced, not money made. I think this is a mistake. People get emotionally connected to money in a way they seldom do with physical product. Looking at a bin of rejected bottles, they may realize on a rational level that they need to do better. If they can look at the bin and see $50, they will connect on a deeper, more emotional level.

A three-minute line stoppage may seem pretty trivial, and most people find it hard to get too excited. A few might see three minutes and realize that 1,200 products (at 400 products per minute) did not get produced. Fewer still will realize that those three minutes represent a loss of $600, assuming 50 cents per product. Short stoppages like this, repeated several times a day, can add up to half a million dollars or more over a year.

Everyone must recognize costs. This info is not just for management; it needs to be disseminated as widely as possible. The operators on the line must know the dollar cost of every product that passes them. The mechanic setting up the machine must know the dollar cost of each minute the machine is not running. The warehouse crew must know the dollar cost of every minute the

Machinery Matters

line is not running because of delays in getting materials to it. Don't keep this information secret.

Mount posters on each line, showing each product, its cost and the dollar cost per minute of stoppages. This will help with awareness.

I saw one graphic way to do on a personal-care line some years ago. A monitor at the end of the line displayed safety and motivational messages. Whenever the line stopped, these were replaced by a dollar sign and a spinning odometer with a big dollar sign. Each second would increment by 90 cents. On restart, it would display the total cost of the stoppage. It really focused people on what was important.

Rejected, discarded or damaged components, such as caps or bottles, while individually cheap, will add up. One plant has bins to collect various types of waste; each bin was labeled with a dollar amount. A bottle bin, when full, represented $90. Making the dollars visible keeps them in people's minds

Your plant is not in business to make product. It is in business to make money. Go look at the end of your line, it really is money, even if it looks like product. A dollar here, a quarter there, and pretty soon you are talking about real money!

This column originally appeared in the February 2010 issue of Food & Beverage Packaging magazine

John R Henry CPP

42 Fast forward: The packaging line of 2019

NOTE: Of all the articles I've written, this was probably the most fun to write. My editor, Lisa Pierce, asked me to prognosticate what packaging lines would look like 10 years in the future. She and I came up with the following.

I always laugh at movies from the 60's showing scientists in the 90's still using slide rules. How could they get it so wrong? Well, making accurate predictions is tough, especially when we try to do it about the future. In 2019, let's see how wrong, or perhaps how right, I was. Every year 3D printing/rapid prototyping technologies become more economical and feasible. I am beginning to wonder if the transmogrifier I mention is that far off the mark.

John R Henry – June 2011

Much of the groundwork has already been laid. We just need to continue building improvements in from there.

The cutting-edge packaging line of the future will be affected by many forces. Some will be internal. Packagers will always be looking to cut costs and packaging material, labor and efficiency is always a possibility. Market forces will drive progress as customers demand more innovative packaging and competitors supply it. A third force is societal. Packaging, mostly wrongly, has a reputation of being wasteful of resources, as well as taking excessive space in landfills. Packagers will need to make packaging greener.

So here is one prediction of what the packaging line of the future will look like:

Machinery Matters

First and foremost, the packaging line of the future will be flexible. The age of mass customization is truly upon us with consumers demanding what they want, how, when and where they want it. Marketing channels, each with different requirements, are proliferating. Traditional outlets must compete with hypermarkets, convenience stores, club stores, dollar stores and boutique food marts, as well as specialty stores and other outlets. A product like Coca-Cola—which was available in a single flavor/size/style as recently as the mid-'60s—is now produced in more than 150 combinations This does not include the many non-Coca-Cola variations such as Sprite and other flavors. The line that seldom if ever needed to introduce a new product may now need to introduce one every month. Flexibility in the line design is required to allow it to accommodate these unknown future products in a timely and cost-effective manner.

This explosion of products has meant the death of the dedicated line. Most packaging lines must be changed over several times a week. In some cases, several times a day. Eight- to 12-hour changeovers are no longer acceptable. Changeovers within 10 minutes will become the goal. Changeover will need to be, as Paul Zepf of Zarpac says: "No time, no tools, no talent, no tinkering." It must be easily performed in a minimal amount of time by semi-skilled operators with the line running at full efficiency immediately at restart. The line of the future will be designed with this concept in mind. Machine builders will finally realize this and design their machines for flexibility. Technology will advance flexibility. In recent years individual servo motors have been replacing the traditional main motor using cams, chains and linkages to drive machine functions. Servos eliminate much mechanism and simplify machine construction and maintenance while improving reliability. They also allow functionality not possible before.

For example, before servos, it was impossible to measure individual cap torque. As servos replace mechanical clutches and drives, they allow the on-torque of every cap to be measured and recorded. More importantly, they allow the torque to be set and controlled via software. This eliminates the need for a mechanic with wrenches, as well as the downtime and inaccuracies inherent in mechanical adjustment. Setting is done by the operator selecting the product name from the control panel.

Servo motors are also being used on cartoner flap tuckers, thermoformer platens, flowrapper metering conveyors and other functions. As they will continue to decrease in price and increase in functionality, by 2019 they will supply all power on packaging machines. As this happens, daily changeover will become fully automatic. New product introductions will be hastened and simplified as modifications will be made by programming rather than replacing parts.

Communications will be standardized. Presently it can be difficult for different machines to communicate with each other as required for smooth line operation. The Organization for Machine Automation and Control (OMAC) is working to establish standards that all machine builders will adhere to. This uniform set of standards will ease maintenance, as technicians will only have to learn a single set of operational protocols. This will result in smoother operation and less downtime.

Changes in machine construction will result in changes to the plant workforce. No longer will there be large numbers of operators and technicians. Machines will become more autonomous, reducing or eliminating the need for semi-skilled operators. Components and product will feed to the line directly from storage with no handling. Finished product will be palletized and placed on an Automatically Guided Vehicle (AGV). The AGV will transport the product to the warehouse or a waiting truck, never being touched by humans.

The line will be autonomous not only in operation but in diagnostics and repair as well. Sophisticated computer control systems will monitor every aspect of operation. Sensors will detect impending problems and either repair them via software commands where possible or alert the technician before line operation can be affected. Most current alarm systems are hopelessly dumb. The line of the future will send a message to the technician's smart phone at the first hint of malfunction. That message will not only announce the problem, it will prioritize it, give details on its nature and advise the materials and tools required to resolve it. When the technician arrives at the line they will not only know exactly what they need to do, they will have everything necessary, eliminating the need to chase back to the tool room.

This puts an almost scary burden on the technician. Although the machine will tell them what to do, they will require a high level of skill to do it. High levels of training will become a necessity. Packagers will demand and machine builders will provide programs to develop these skills. The traditional manual will still be available but supplemented by audio visual versions at the machine controller as well as on pocket-size wireless devices. These will allow every technician to have every scrap of information available about the machine including diagrams, parts lists and more in their pocket, wherever they may be.

Or else:

By 2019 quantum physics, nanotechnology and high-level electronics will have intersected to eliminate packaging altogether. Everyone will have their own personal transmogrifier. This will allow the consumer to simply imagine the product that they want and the transmogrifier will produce it out of thin air.

Check back in 10 years to see which vision is more accurate.

Prototype Packaging Line of 2019
(By Lisa Pierce)

Product Name: Gulp Pulp

Description: A delicious, nutritious puree of vegetables, proteins and fruit for use as a sauce, topping or refreshing ready-to-consume snack

Flavors: Strawberry Sublime, Blueberry Bliss and Kiwi Kismet

Sizes: 12, 16, 24 and 32 fluid ounces

Refrigerate after opening.

1. **Order is automatically generated** from retail store or consumer website and directed to closest regional manufacturing center for direct-store-delivery or direct-consumer-delivery. Enterprise Resource Planning (ERP) software interfaces with line controls throughout the process.

2. Product is formulated from pre-cooked ingredients and pumped to packaging lines on-demand. Red pipe conveys Strawberry Sublime; blue, Blueberry Bliss; green, Kiwi Kismet.

3. Semi-rigid, mono-material rollstock—preprinted with generic brand graphics—is brought to the aseptic filling machines by next-generation vision-controlled, robotic, automated guided vehicles (AGVs). Each filler accommodates two rolls and splicing is done automatically for continuous operation.

4. High-quality, wide-web ink jet printer adds product-specific information, as well as prints random electronic code with conductive ink on each package.

5. Ink for ink jet printers is automatically replenished through overhead pipes.

6. AGVs also automatically refill hoppers with easy-open and reclosable metering dispensing closures, which are fed to each filler, sterilized and sealed onto rollstock prior to forming, filling and final sealing.

7. X-ray unit inspects each pack as it exits filler in single file. Damaged, underweight or otherwise imperfect packages are rejected and sent to holding bin for manual visual inspection.

8. Underweight packages are accumulated, manually packed into cases and donated to charity.

9. Unsalables are automatically emptied. Product is bulk-packed and sold for use as soil nutrients to local farmers. Packages and dispensers are automatically separated and crated for recycling or in-plant waste-to-energy incineration.

10. If needed, first-in/first-out accumulating conveyor holds up to six minutes of packages.

11. Packs enter multipacking station and are collated/organized by pick-and-place robots into same-flavor or mixed-flavor 3x3 cubes, which are then wrapped with a clear film that's specially treated to prevent any horizontal movement when multipacks are vertically stacked.

12. Consumer orders exit left, and a mailing label is immediately printed and applied. Packs are shipped via commercial carrier. Multiple stations can be set up as needed when direct-to-consumer deliveries increase.

13. If needed, first-in/first-out accumulating conveyor holds up to six minutes of multipacks.

14. Robotic unitizers hold hundreds of patterns. Each robot automatically reads electronic product codes and picks appropriate multipacks from the looped conveyor for the specific order. Each unitized load is delivered directly to the store on-demand, so standard slip sheets now measure 20 x 24 inches. Unitized loads are double-stacked prior to being automatically loaded into waiting trucks.

15. Traversing printer (in three dimensions) codes customer-specific order data on unitized-load label (human- and machine-readable), and applicator wraps labels around opposite corners so all sides are tagged. Ink is piped to printers from stored bulk supply, and labels are automatically replenished by AGVs. Individual package graphics create high-impact billboards for end-aisle or club-store displays.

16. The 24/7 lights-out facility requires just one employee to monitor the lines. Section lighting turns on when it senses the person's photo ID is within a 10-yard area. Any alerts are automatically sent to the operator's wireless device with Bluetooth headset (wireless service is on a secured network) with specific problem and solution(s) identified. Otherwise, the worker is free to listen to audio of his/her choice.

This column originally appeared in the April 2009 issue of Food & Beverage Packaging magazine

About the author

John Henry is the Changeover Wizard at www.changeover.com He assists packagers and other manufacturers improve their operational efficiencies.

John has also been a contributing writer for Food & Beverage Packaging magazine since 2001. In addition to articles in F&BP, he has published numerous articles in other journals and magazines.

John developed and presents the Institute of Packaging Professionals' (IOPP) twice yearly course, Packaging Machinery Basics. He also wrote the textbook Fundamentals of Packaging Machinery for the IOPP.

John is a longtime and active member of the IOPP and a lifetime Certified Packaging Professional (CPP).

Learn more at his website: www.changeover.com

CPSIA information can be obtained at www.ICGtesting.com
Printed in the USA
BVOW021435210113

311192BV00018B/887/P